The Science of
DNA and Evolution
From Molecules to Miracles

ELIA M TRURES

Copyright © 2025 ELIA M TRURES

All rights reserved.

The Science of DNA and Evolution From Molecules to Miracles

Table of Contents

Part 1: The Blueprint of Life

Page 5 to 15

Part 2: Evolution and Natural Selection

Page 15 to 34

Part 3: The Power of DNA in Science and Medicine

Page 34 to 58

Part 4: DNA and the Future of Life

Page 58 to

Conclusion: The Code of Life and Our Place in the Universe

Page 80 to 101

Part 1: The Blueprint of Life

Introduction:

The Mystery of Life's Code

For centuries, humanity has sought to understand the fundamental question: What makes life possible? The answer lies in an elegant and intricate molecule—DNA. This double-helixed strand, hidden within the nucleus of every living cell, carries the instructions for building and sustaining life itself.

The discovery of DNA's structure in 1953 by James Watson and Francis Crick, building on the work of Rosalind Franklin and Maurice Wilkins, revolutionized our understanding of biology. It unveiled a molecular script written in a four-letter alphabet—adenine (A), thymine (T), guanine (G), and cytosine (C)—that dictates the development, function, and evolution of every organism on Earth.

But DNA is more than just a biological blueprint; it is a code that connects all living beings, from bacteria to humans, tracing an unbroken thread through billions of years of evolution. This genetic language, passed down through generations, allows life to adapt, survive, and thrive in an ever-changing world.

Despite these breakthroughs, the full power of DNA remains one of the greatest scientific mysteries. How did this code emerge? How does it influence traits, behavior, and even disease? Can we rewrite life's code to cure genetic disorders or extend human longevity? These questions drive modern research in genetics, biotechnology, and medicine, bringing us closer to unlocking the deepest secrets of life.

As we embark on this journey into the world of DNA, evolution, and genetic engineering, we uncover not only the story of life's origins but also the potential to shape its future. The mystery of life's code is one of the most profound and exciting frontiers of science—one that continues to evolve with every new discovery.

What is DNA? – The Molecular Basis of Life

DNA (Deoxyribonucleic Acid) is the fundamental blueprint of life. Found in nearly every living cell, it carries the genetic instructions that determine an organism's growth, development, function, and reproduction. Acting as a biological code, DNA stores and transmits hereditary information from one generation to the next, ensuring continuity of life across billions of years of evolution.

The Structure of DNA: The Double Helix

The structure of DNA was famously discovered in 1953 by James Watson and Francis Crick, with critical contributions from Rosalind Franklin and Maurice Wilkins. They revealed that DNA is a **double helix**, resembling a twisted ladder. This structure is composed of:

Nucleotides: The basic building blocks of DNA, consisting of:

A **phosphate group**, A **deoxyribose sugar**, A **nitrogenous base**

The Four Bases: DNA's genetic code is written using four nitrogenous bases:

Adenine (A), Thymine (T)

Guanine (G), Cytosine (C)

These bases pair specifically — A always pairs with T, and G always pairs with C — forming the rungs of the DNA ladder. This complementary base pairing is crucial for DNA replication and genetic stability.

DNA's Role in Life

DNA is essential for all cellular functions and hereditary processes. Its primary roles include:

Storing Genetic Information: DNA contains genes, which are segments of genetic code that instruct cells on how to produce proteins.

Replication: Before a cell divides, DNA is copied so that each new cell receives an identical set of genetic instructions.

Protein Synthesis: Through processes called **transcription** and **translation**, DNA directs the production of proteins, which are essential for an organism's structure and function.

Genetic Variation and Evolution: Mutations in DNA contribute to genetic diversity, enabling species to adapt and evolve over time.

The Significance of DNA in Modern Science

The discovery of DNA's structure and function has revolutionized many fields, from medicine to forensic science. Understanding DNA has led to advancements such as:

Genetic engineering and **CRISPR** technology for gene editing

DNA sequencing to identify genetic disorders

Forensic DNA analysis for solving crimes

Personalized medicine based on genetic profiles

DNA is more than just a molecule — it is the very essence of life itself. From the smallest bacteria to the complexity of the human body, this remarkable code continues to shape and define the biological world. As research advances, DNA remains at the heart of some of the greatest scientific discoveries of our time.

The Discovery of DNA – From Mendel to Watson and Crick

The journey to understanding DNA, the molecule that carries genetic information, spans centuries and involves the work of many brilliant scientists. From Gregor Mendel's early experiments on heredity to the groundbreaking discovery of DNA's double-helix structure by James Watson and Francis Crick, the story of DNA is one of curiosity, perseverance, and scientific revolution.

Gregor Mendel – The Father of Genetics

In the mid-nineteenth century, an Austrian monk named Gregor Mendel conducted experiments on pea plants in his monastery garden. Through careful observation, he discovered patterns of inheritance, which led to the formulation of Mendel's Laws of Inheritance. His work demonstrated that traits are passed from parents to offspring through discrete units, later known as genes. Though his research was largely ignored at the time, it laid the foundation for the field of genetics.

Friedrich Miescher – The First Discovery of DNA

In 1869, Swiss chemist Friedrich Miescher was the first to identify a new substance within the nuclei of white blood cells, which he called "nuclein." This substance was later recognized as DNA. However, at the time, scientists believed that proteins, not DNA, were responsible for heredity.

Oswald Avery and the Proof That DNA is Genetic Material

For many years, scientists debated whether proteins or DNA carried genetic information. In 1944, Oswald Avery and his colleagues, Colin MacLeod and Maclyn McCarty, conducted experiments demonstrating that DNA, not protein, was the substance responsible for heredity. This was a crucial step in confirming the role of DNA in genetics.

Erwin Chargaff – The Base Pairing Rules

In the 1950s, biochemist Erwin Chargaff analyzed the composition of DNA and found that the amount of adenine always matched the amount of thymine, and the amount of guanine always matched the amount of cytosine. This discovery, known as Chargaff's Rules, provided key insights into the structure of DNA.

Rosalind Franklin and Maurice Wilkins – The X-ray Clue

Using a technique called X-ray crystallography, Rosalind Franklin captured an image of DNA, known as **Photo 51**, which revealed its helical structure. Her colleague, Maurice Wilkins, later shared this image with Watson and Crick without her direct permission, which played a crucial role in their breakthrough discovery.

James Watson and Francis Crick – The Double Helix

In 1953, James Watson and Francis Crick used Franklin's X-ray data along with Chargaff's base pairing rules to build the first accurate model of DNA's structure. They proposed that DNA is a **double helix**, with two strands twisted around each other, held together by complementary base pairs (adenine pairing with thymine and guanine pairing with cytosine). This discovery revolutionized biology, explaining how DNA replicates and how genetic information is passed from one generation to the next.

The Legacy of DNA's Discovery

The discovery of DNA's structure marked the beginning of modern genetics and molecular biology. It paved the way for advances such as DNA sequencing, genetic engineering, forensic DNA analysis, and medical breakthroughs in treating genetic diseases. Today, our understanding of DNA continues to evolve, shaping the future of science and medicine in ways once thought impossible.

The story of DNA is a testament to the power of scientific inquiry, with each discovery building upon the work of those before. From Mendel's simple pea plant experiments to the double-helix model, the unraveling of life's code remains one of the greatest achievements in science.

Genes and the Genome – Mapping Life's Instructions

Every living organism carries a unique set of instructions that determine its traits, functions, and development. These instructions are encoded in **genes**, which together form the **genome** — the complete set of genetic material within an organism. Understanding genes and the genome has revolutionized biology, medicine, and biotechnology, providing insights into heredity, evolution, and disease.

What Are Genes?

A **gene** is a segment of DNA that contains the instructions for making a specific **protein**. Proteins are essential molecules that perform countless functions, from building tissues to regulating biochemical reactions. Each gene acts like a coded message, with its sequence of **nucleotides** (adenine, thymine, guanine, and cytosine) dictating how proteins are assembled.

Genes vary in size and complexity. Some genes control simple traits like eye color, while others are involved in complex processes such as metabolism, immunity, and brain function. Though humans have approximately **20,000–25,000 genes**, they account for only a small fraction of the entire genome.

What Is the Genome?

The **genome** is the complete set of an organism's DNA, including all its genes and non-coding regions. It contains all the information necessary for growth, development, and survival. The **human genome**, for example, consists of around **3 billion base pairs** organized into **23 pairs of chromosomes**.

Non-coding regions of the genome, once thought to be "junk DNA," are now recognized for playing crucial roles in **gene regulation**, **chromosome stability**, and **evolutionary processes**. These regions help control when and how genes are activated or silenced, influencing everything from development to disease susceptibility.

Mapping the Genome – The Human Genome Project

One of the most groundbreaking scientific achievements was the Human Genome Project (HGP), an international research effort launched in 1990 to sequence the entire human genome. Completed in 2003, the project provided an essential blueprint of human DNA, revealing the precise locations of genes and variations across individuals.

The HGP has led to numerous advances, including:

Genetic testing for inherited diseases

Personalized medicine, tailoring treatments to an individual's genetic makeup

Gene therapy, offering potential cures for genetic disorders

Evolutionary studies, comparing genomes across species to understand human origins

Genes, Mutations, and Evolution

Changes in genes, known as **mutations**, can have profound effects. Some mutations are **beneficial**, driving **evolutionary adaptation**, while others cause **genetic disorders** like sickle cell anemia or cystic fibrosis. Understanding genetic mutations has been key to studying evolution, biodiversity, and the origins of life.

The Future of Genomics

Advances in genome sequencing and **CRISPR** gene-editing technology are transforming science and medicine. Scientists can now edit genes to correct mutations, potentially curing diseases at the genetic level. Research into the **epigenome**, which studies how environmental factors influence gene expression, is also shedding light on how lifestyle choices impact health.

Mapping life's instructions through genes and the genome has opened a new frontier in science. From understanding hereditary diseases to unlocking the secrets of human evolution, genetics continues to shape the future of biology, medicine, and beyond.

How DNA Works – Transcription, Translation, and Protein Synthesis

DNA serves as the master blueprint for life, containing instructions for building and maintaining an organism. However, DNA itself does not perform any direct functions in the cell. Instead, it relies on a complex process called **gene expression** to convert genetic information into functional proteins. This process occurs in two key stages: **transcription** and **translation**, together known as **protein synthesis**.

1. Transcription – From DNA to RNA

The first step in gene expression is **transcription**, where a gene's DNA sequence is copied into a molecule of **messenger RNA (mRNA)**. This occurs in the **nucleus** of eukaryotic cells (or the cytoplasm in prokaryotes) and is carried out by an enzyme called **RNA polymerase**.

Steps of Transcription:

Initiation
RNA polymerase binds to a specific region of DNA called the **promoter**, which signals the start of a gene.

The DNA strands unwind, exposing the template strand.

Elongation
RNA polymerase moves along the DNA, reading the template strand and assembling a complementary strand of **mRNA** using **RNA nucleotides** (adenine, uracil, cytosine, and guanine).

In RNA, **uracil (U)** replaces **thymine (T)** found in DNA.

Termination
Once RNA polymerase reaches a termination signal, transcription stops, and the newly formed **mRNA** detaches.

In eukaryotic cells, the mRNA undergoes further processing, including the addition of a **5' cap**, a **poly-A tail**, and the removal of non-coding regions (**introns**) through **splicing**.

2. Translation – From mRNA to Protein

After transcription, the **mRNA** carries genetic instructions from the nucleus to the **ribosome**, the site of protein synthesis in the cytoplasm. This stage is called **translation** because the nucleotide sequence of mRNA is "translated" into an amino acid sequence, forming a protein.

Key Players in Translation:

mRNA: Carries the genetic code from DNA.

Ribosome: Reads the mRNA sequence and assembles amino acids into a protein.

Transfer RNA (tRNA): Matches amino acids to the corresponding mRNA **codons** (three-nucleotide sequences).

Amino acids: The building blocks of proteins, linked together to form polypeptides.

Steps of Translation:

Initiation

The ribosome binds to the mRNA at the **start codon (AUG)**, which signals the beginning of protein synthesis.

A **tRNA** carrying the amino acid **methionine** attaches to the start codon.

Elongation

The ribosome moves along the mRNA, reading codons.

Each codon is matched with a complementary **tRNA** carrying the appropriate **amino acid**.

The ribosome catalyzes the formation of **peptide bonds**, linking amino acids into a growing polypeptide chain.

Termination

When the ribosome reaches a **stop codon (UAA, UAG, or UGA)**, translation ends.

The newly synthesized protein is released and undergoes **folding** and **modifications** efore becoming functional.

3. The Significance of Protein Synthesis

Protein synthesis is essential for all cellular functions, as proteins serve various roles, including:

Structural proteins (collagen, keratin) that support tissues.

Enzymes (DNA polymerase, amylase) that catalyze biochemical reactions.

Hormones (insulin, growth hormone) that regulate body functions.

Antibodies that help the immune system fight infections.

By controlling **transcription and translation**, cells regulate which proteins are produced, allowing them to adapt to environmental changes and perform specialized functions.

Understanding the mechanisms of gene expression has led to breakthroughs in **genetic engineering**, **gene therapy**, and **biotechnology**, shaping the future of medicine and science.

Part 2: Evolution and Natural Selection

Darwin's Legacy – The Theory of Evolution

The theory of evolution is one of the most influential ideas in science, fundamentally changing our understanding of life on Earth. Proposed by Charles Darwin in the 19th century, the theory explains how species change over time through natural selection. Today, Darwin's ideas remain central to biology, influencing fields from genetics to medicine and conservation.

1. Charles Darwin and the Birth of Evolutionary Theory

Charles Darwin, an English naturalist, developed his theory of evolution after a five-year voyage on the HMS Beagle (1831–1836). During his journey, Darwin observed diverse species and their adaptations, particularly in the Galápagos Islands. He noticed that finches on different islands had varying beak shapes suited to their diets, suggesting that species evolve to better survive in their environments.

In 1859, Darwin published *On the Origin of Species*, where he outlined the principles of evolution by **natural selection**. His work was groundbreaking because it provided a scientific explanation for how life diversifies over generations.

2. Natural Selection – The Driving Force of Evolution

Natural selection is the process by which certain traits become more common in a population because they provide a survival or reproductive advantage. The key principles of natural selection are:

Variation: Individuals in a population have differences in traits, such as size, color, or strength.

Competition: Organisms compete for limited resources like food, mates, and shelter.

Survival of the Fittest: Individuals with beneficial traits are more likely to survive and reproduce.

Inheritance: Successful traits are passed on to the next generation, gradually changing the species over time.

Over millions of years, natural selection leads to the development of new species, a process known as **speciation**.

3. Evidence for Evolution

Since Darwin's time, a vast amount of evidence has confirmed the theory of evolution, including:

Fossil Record

Fossils provide snapshots of ancient life, showing how species have changed over time. Transitional fossils, like *Archaeopteryx* (a link between reptiles and birds), demonstrate evolutionary connections.

Comparative Anatomy

Similar structures in different species, known as **homologous structures**, indicate common ancestry. For example, the forelimbs of humans, bats, and whales share the same bone structure despite different functions.

Genetics and DNA

Modern genetics has strengthened Darwin's theory. The discovery of DNA showed that all living organisms share a common genetic code, providing direct evidence of evolutionary relationships.

Embryology

Early embryos of different animals, including humans, fish, and reptiles, look remarkably similar, suggesting shared ancestry.

Biogeography

The distribution of species across the world supports evolution. For example, unique species found in isolated places like Australia and the Galápagos evolved separately from their relatives on other continents.

4. Evolution in Action

Evolution is not just a theory of the past—it is happening today. Examples include:

Antibiotic Resistance: Bacteria evolve resistance to antibiotics, making some infections harder to treat.

Pesticide Resistance: Insects evolve resistance to pesticides, requiring stronger or new chemicals.

Artificial Selection: Humans have guided evolution through selective breeding, producing domesticated animals like dogs and agricultural crops like corn.

5. The Impact of Darwin's Legacy

Darwin's theory has had profound effects on science and society. It has led to advances in **genetics, medicine, ecology, and conservation**, helping us understand biodiversity and protect endangered species. While initially controversial, evolution is now a cornerstone of modern biology, supported by overwhelming scientific evidence.

By uncovering the process that shapes life, Darwin transformed our understanding of nature and our place in it. His legacy continues to inspire new discoveries, shaping the future of evolutionary science.

Mutations and Genetic Variation – The Engine of Evolution

At the heart of **evolution** lies **genetic variation**, the diversity of genes within a population. This variation provides the raw material for **natural selection**, allowing species to adapt and evolve over time. The primary source of genetic variation is **mutations**, which introduce new traits into a gene pool. Without mutations, evolution would not be possible.

1. What Are Mutations?

A **mutation** is a change in the **DNA sequence** of an organism. Mutations can occur naturally during **DNA replication** or be caused by external factors like radiation, chemicals, or viruses. They can affect a single **nucleotide** (point mutation) or involve large segments of chromosomes.

Types of Mutations:

Point Mutations – A single nucleotide is changed, inserted, or deleted.

Silent Mutation: No effect on the protein.

Missense Mutation: Changes one amino acid in a protein (e.g., sickle cell anemia).

Nonsense Mutation: Creates a stop signal, leading to a shortened protein.

Insertion/Deletion Mutations – Extra nucleotides are added or removed, potentially shifting the reading frame (**frameshift mutation**), which can severely impact protein function.

Chromosomal Mutations – Large-scale changes such as duplications, deletions, inversions, or translocations of DNA segments.

2. How Mutations Create Genetic Variation

Mutations introduce new **alleles** (different versions of a gene) into a population. If these mutations occur in reproductive cells (sperm or egg), they can be inherited by future generations. Over time, beneficial mutations may spread through populations, leading to adaptation.

However, not all mutations are beneficial:

Beneficial mutations provide an advantage (e.g., antibiotic resistance in bacteria).

Neutral mutations have no immediate effect but may become important later.

Harmful mutations can cause genetic disorders or reduce an organism's fitness.

3. Other Sources of Genetic Variation

While mutations are the primary source, genetic variation also arises from:

Recombination (Genetic Shuffling)

During **meiosis**, homologous chromosomes exchange genetic material in a process called **crossing over**, creating new combinations of genes.

Gene Flow (Migration)

When individuals move between populations and interbreed, they introduce new genes, increasing genetic diversity.

Genetic Drift

In small populations, random changes in allele frequency can occur due to chance events, sometimes leading to loss of genetic diversity.

4. Mutations and Evolutionary Change

Mutations fuel **evolution** by providing the diversity needed for **natural selection** to act upon. Examples of mutation-driven evolution include:

Antibiotic Resistance in Bacteria – Random mutations can make bacteria resistant to drugs, leading to the evolution of superbugs.

Lactose Tolerance in Humans – A mutation allowed some human populations to digest lactose into adulthood, an adaptation to dairy farming.

Peppered Moth Evolution – During the Industrial Revolution, a mutation led to darker moths, which had a survival advantage in polluted environments.

5. The Role of Mutations in Medicine and Biotechnology

Understanding mutations has led to breakthroughs in **genetics, medicine, and biotechnology**, including:

Cancer Research – Studying mutations in genes that control cell division helps in developing targeted therapies.

Gene Therapy – Scientists can correct genetic disorders by editing or replacing faulty genes.

CRISPR Gene Editing – A revolutionary tool that allows precise modification of DNA to treat diseases or improve crops.

Mutations and genetic variation are the foundation of evolution. While some mutations cause harm, others drive **adaptation, innovation, and survival**. By studying these changes, scientists can better understand evolution, fight genetic diseases, and shape the future of medicine and biotechnology.

Natural Selection in Action – Survival of the Fittest

Natural selection is the primary mechanism of evolution, famously described by Charles Darwin as "**survival of the fittest**." It explains how organisms with traits that better suit their environment tend to survive and reproduce more successfully, passing those advantageous traits to future generations. Over time, natural selection shapes species to adapt to changing environments, driving evolutionary change.

1. What Is Natural Selection?

Natural selection works through the following steps:

Variation: Within any population, individuals exhibit genetic variation. This variation is often caused by **mutations**, **genetic recombination**, or **gene flow**.

Competition: Organisms compete for limited resources like food, mates, and shelter.

Survival of the Fittest: Individuals with traits better suited to the environment have a higher chance of surviving and reproducing.

Inheritance: Beneficial traits are passed on to offspring, gradually becoming more common in the population.

The term **"fitness"** refers not only to an organism's strength but its ability to survive, reproduce, and pass on its genes to the next generation.

2. Examples of Natural Selection in Action

Natural selection can be observed in numerous examples across nature, from bacteria to mammals. Some of the most famous examples include:

1. Peppered Moths

During the Industrial Revolution, the environment around England's cities became darker due to pollution. In this polluted environment, darker-colored **peppered moths** (Biston betularia) became more camouflaged against the soot-covered trees, while lighter-colored moths were more easily spotted by predators. Over time, the population shifted, and the dark-colored moths became more common. This is a clear example of **directional selection**, where one phenotype (dark moths) became favored over another (light moths).

2. Darwin's Finches

On the Galápagos Islands, Darwin observed a variety of finches with different beak shapes, each adapted to specific food sources. During periods of drought, finches with larger beaks, capable of cracking hard seeds, had a better chance of surviving and reproducing. When conditions changed, finches with smaller beaks, suited for eating softer seeds, flourished. This represents **disruptive selection**, where different traits in a population are favored in different environmental conditions.

3. Antibiotic Resistance in Bacteria

In bacteria, natural selection can occur rapidly. When antibiotics are used, most bacteria may be killed, but some, due to mutations, may possess resistance to the antibiotic. These resistant bacteria survive, reproduce, and pass on the resistance genes. Over time, this leads to the development of **superbugs**, which are bacteria that can no longer be treated by conventional antibiotics. This is an example of **selective pressure** causing the rapid evolution of traits in response to human actions.

4. Giraffe Neck Evolution

One of the classic examples is the evolution of the giraffe's long neck. It is believed that during periods of scarce food, giraffes with longer necks could reach higher branches of trees, giving them a feeding advantage. Over time, natural selection favored longer-necked giraffes, leading to the species we know today. This is an example of **sexual selection**, where the trait of longer necks also played a role in attracting mates, further reinforcing the advantageous trait.

3. Types of Natural Selection

Natural selection can occur in several forms, depending on how environmental pressures affect a population:

Directional Selection: Favors one extreme phenotype (e.g., dark moths during industrial pollution).

Stabilizing Selection: Favors the average or intermediate phenotype and selects against extremes. For example, babies born within a certain weight range tend to have higher survival rates, while very small or very large babies may face challenges.

Disruptive Selection: Favors extreme phenotypes at both ends of the spectrum, often leading to two distinct groups within the population (e.g., large and small beaked finches).

Sexual Selection: A form of selection where traits that improve an organism's chances of attracting a mate become more common, even if they don't directly improve survival. Examples include the colorful tail feathers of male peacocks or the elaborate courtship rituals of certain birds.

4. The Future of Natural Selection and Evolution

In today's rapidly changing world, natural selection continues to shape life on Earth. However, human actions have introduced new factors influencing evolution, such as:

Climate change, which alters habitats and creates new selective pressures.

Habitat destruction, which can reduce genetic diversity and change the direction of selection.

Human intervention in the form of medicine, agriculture, and conservation efforts, which can artificially change the evolutionary pressures on species.

Understanding natural selection helps us address challenges like **disease control**, **biodiversity conservation**, and **climate adaptation**. It also helps scientists predict how species may evolve in response to changing environmental factors.

Natural selection is a powerful and ongoing process that drives evolution. Through **survival of the fittest**, organisms with advantageous traits have a better chance of surviving and reproducing, ensuring the continuation and adaptation of species. Whether it's the evolution of **antibiotic resistance** or the survival of the **dark peppered moths**, natural selection remains a fundamental principle shaping life on Earth.

Speciation and Adaptation – How New Species Emerge

Speciation is the process by which new species arise from a common ancestor. This is a fundamental concept in evolutionary biology, explaining how the diversity of life on Earth has expanded and continues to change over time. It is often driven by **adaptation** to different environments and the mechanisms of **natural selection**. Understanding how new species emerge helps us grasp the complexity of life and the ongoing process of evolution.

1. What is Speciation?

Speciation occurs when populations of a single species become so genetically different that they can no longer interbreed and produce fertile offspring. Over time, these populations evolve into distinct species. Speciation can happen in several ways, but the most common forms are **allopatric speciation, sympatric speciation**, and **parapatric speciation**.

2. Types of Speciation

1. Allopatric Speciation (Geographic Isolation)

This is the most well-known and common form of speciation. Allopatric speciation occurs when a population is geographically divided into two or more isolated groups by physical barriers such as mountains, rivers, or human-made structures. Over time, these isolated groups experience different selective pressures, mutations, and genetic drift, leading to distinct evolutionary paths.

Example: The **Galápagos finches** studied by Darwin evolved into multiple species after being isolated on different islands, each with its own environmental conditions. Each population developed distinct beak shapes and sizes adapted to different food sources on their respective islands.

2. Sympatric Speciation (Same Location)

In contrast to allopatric speciation, sympatric speciation occurs without geographic isolation. Populations within the same geographic area become reproductively isolated due to other factors such as **behavioral differences**, **dietary preferences**, or **chromosomal changes**.

Example: In some fish species, like **cichlids** in Lake Victoria, differences in mating preferences or habitat use can lead to the development of reproductive isolation, even though they live in the same lake. This process can occur rapidly, particularly in environments with many ecological niches.

3. Parapatric Speciation (Partial Isolation)

Parapatric speciation occurs when populations are partially isolated but still have some overlap in their geographic ranges. There is limited gene flow between the populations, and environmental gradients (changes in habitat or climate) drive the evolution of distinct species.

Example: The **grass species Anthoxanthum odoratum** has populations that are partially isolated due to environmental factors such as soil type. Different selective pressures in different areas cause the populations to evolve in different directions, potentially leading to speciation over time.

3. The Role of Adaptation in Speciation

Adaptation is a key driver of speciation. As populations encounter different environments, they undergo genetic changes through **mutation, genetic recombination**, and **natural selection**. Over time, these adaptations become pronounced, leading to differences in traits such as **morphology, behavior**, and **physiology**. These differences can eventually result in reproductive isolation, which is essential for speciation.

Examples of Adaptation Leading to Speciation

Darwin's Finches

On the Galápagos Islands, different finch populations adapted to different food sources. Some evolved larger beaks for cracking hard seeds, while others developed smaller beaks for eating soft seeds. Over time, these adaptations led to the formation of multiple finch species, each suited to a specific ecological niche.

Cichlid Fish in African Lakes

The cichlid fish in Lake Tanganyika and Lake Victoria in Africa exhibit remarkable diversity, with hundreds of species having evolved in response to different ecological niches, such as variations in water depth, diet, and territorial behavior. This is an example of how **adaptive radiation** (the rapid diversification of a species into many different forms) leads to speciation in a single environment.

Polar Bears and Brown Bears

The **polar bear** evolved from brown bears that adapted to cold environments. Over time, through natural selection, polar bears became well-suited to life in the Arctic, with thick fur, large paws for walking on ice, and a diet primarily consisting of seals. This adaptation to extreme cold, along with genetic divergence, led to speciation.

4. Reproductive Isolation – The Key to Speciation

For speciation to be complete, populations must become reproductively isolated. This means that even if individuals from different populations meet, they cannot successfully mate and produce fertile offspring. Reproductive isolation can be caused by several mechanisms:

1. Prezygotic Barriers (Before Fertilization)

These prevent mating or fertilization between species and include:

Behavioral Isolation: Differences in mating rituals or behaviors prevent mating (e.g., different songs in birds).

Temporal Isolation: Species reproduce at different times (e.g., different flowering times in plants).

Mechanical Isolation: Structural differences prevent successful mating (e.g., size differences in insect species).

Gametic Isolation: Even if mating occurs, the sperm and egg cannot fuse (e.g., in some marine organisms).

2. Postzygotic Barriers (After Fertilization)

These occur after fertilization and affect the viability or fertility of offspring:

Hybrid Inviability: The hybrid offspring do not survive or develop properly.

Hybrid Sterility: The hybrid offspring are sterile, like the **mule** (a cross between a horse and a donkey).

5. Adaptive Radiation – The Rapid Evolution of New Species

When new environments become available or when species colonize new habitats, **adaptive radiation** can occur. This is a form of speciation where a single ancestral species rapidly diversifies into a wide variety of new species, each adapted to a different ecological niche.

Example: The **Darwin's finches** in the Galápagos are an example of adaptive radiation. After arriving on the islands, a single species evolved into multiple species, each adapted to different food sources and ecological niches.

Example: The **Hawaiian honeycreepers**, a group of birds, underwent adaptive radiation as they adapted to different diets (nectar, insects, seeds) and ecological roles on the Hawaiian Islands.

Speciation is a gradual and ongoing process, driven by **adaptation** to new environments, the emergence of reproductive isolation, and the action of **natural selection**. Whether through geographic isolation, ecological differentiation, or other factors, new species continually emerge as life evolves. Understanding speciation not only provides insight into the history of life but also helps us understand the mechanisms shaping biodiversity on Earth today.

The Fossil Record and Evolutionary Evidence

The fossil record is one of the most compelling pieces of evidence for the theory of evolution. Fossils provide a **historical record** of life on Earth, showing how species have changed over time. They offer direct evidence of past organisms, their characteristics, and how they are related to modern species. The study of fossils, known as **paleontology**, has revealed an astonishing story of life's **diversity** and **adaptation** through millions of years.

1. What is the Fossil Record?

The fossil record refers to the preserved remains or traces of ancient organisms, typically found in **sedimentary rocks**. Fossils can include hard parts like **bones, teeth,** and **shells**, as well as softer structures like **imprints, footprints,** and **burrows**. These fossils are distributed across different layers of rock, known as **strata**, which reflect different time periods in Earth's history.

Fossils are created under specific conditions:

The organism must be buried quickly after death to prevent decay and scavenging.

Over time, minerals replace the organic material, turning it into rock.

The deeper the strata, the older the fossils within them, providing a timeline for Earth's evolutionary history.

2. Fossil Evidence for Evolution

The fossil record offers significant evidence for evolution in several key ways:

1. Transitional Fossils (Missing Links)

Transitional fossils are fossils that show intermediate characteristics between different groups of organisms. These fossils are crucial because they help demonstrate how major evolutionary transitions occurred, providing insight into the evolution of complex structures and behaviors.

Example: The **transition from fish to amphibians** is illustrated by fossils like **Tiktaalik**. This ancient fish had features of both fish (scales, fins) and amphibians (limb-like structures capable of supporting the body on land).

Example: The evolution of **birds from dinosaurs** is evidenced by fossils like **Archaeopteryx**, which had features of both theropod dinosaurs (teeth, claws) and modern birds (feathers, wings).

2. Gradual Changes Over Time

The fossil record shows a pattern of gradual changes in species over millions of years, providing evidence for the process of **natural selection** and **adaptive evolution**.

Example: The evolution of **whales** can be traced through fossils that show gradual changes in limb structure and body shape, such as the transition from land-dwelling mammals (like **Ambulocetus**) to fully aquatic species (like modern **baleen whales**).

3. Fossilized Evidence of Extinction

The fossil record also shows that many species that once thrived are no longer present on Earth. Extinction is a natural part of evolution, and it often paves the way for new species to emerge and diversify.

Example: The **dinosaurs**, which dominated Earth for millions of years, went extinct around 66 million years ago, likely due to a combination of environmental changes, including the impact of a large asteroid. This extinction event paved the way for the rise of **mammals** and the eventual appearance of humans.

4. The Distribution of Fossils and Biogeography

The distribution of fossils across different geographic regions also supports the theory of evolution. Fossils found in regions that are now separated by oceans or mountain ranges show how species evolved and dispersed across Earth's continents over time.

Example: The fossils of **marsupials** in Australia are distinct from those found on other continents. This is evidence of the **continental drift**, the slow movement of Earth's continents, and the resulting **evolutionary divergence** of species in isolated environments.

3. Types of Fossils

There are several types of fossils that provide different kinds of evidence for evolution:

1. Body Fossils

These include the remains of an organism's body, such as bones, teeth, shells, or leaves. They provide direct evidence of the physical characteristics of past life forms.

Example: The **saber-toothed cat** fossils reveal large canine teeth adapted for hunting large prey.

2. Trace Fossils

These include footprints, burrows, and other marks left by organisms. Trace fossils provide insight into the behavior and movement of ancient creatures.

Example: Dinosaur **footprints** show how these animals moved and interacted with their environment, providing evidence of their size, speed, and social behaviors.

3. Amber Fossils

Some organisms have been preserved in tree resin, which later hardened into amber. These fossils can include entire organisms, such as insects, spiders, and even small vertebrates, preserved in incredible detail.

Example: Insects trapped in amber are often perfectly preserved, providing detailed insights into ancient ecosystems.

4. Radiometric Dating – Determining Fossil Age

Radiometric dating techniques allow scientists to determine the age of fossils and the rocks surrounding them. By measuring the decay of radioactive isotopes, such as **carbon-14** or **uranium-238**, scientists can estimate the age of fossils with great accuracy.

Carbon-14 dating is used for relatively recent fossils (up to about 50,000 years old), while other isotopes are used for older fossils.

This allows paleontologists to construct a **timeline** of evolutionary events and understand how long it took for different species to evolve and go extinct.

5. Fossil Evidence for Human Evolution

Fossils have also provided a detailed record of human evolution. By studying the fossilized remains of early humans and their ancestors, scientists have been able to trace the development of human traits like **bipedalism** (walking on two legs), **tool use**, and **brain size**.

Example: Fossils like **Australopithecus afarensis** (e.g., **Lucy**) show evidence of bipedalism, while later fossils like **Homo habilis** and **Homo erectus** display an increase in brain size and the development of early tools. The evolution of humans from primate ancestors is one of the most well-documented examples of evolution through the fossil record.

The fossil record provides overwhelming evidence for the process of **evolution**. It shows how life on Earth has changed over time through gradual changes, the emergence of new species, and the extinction of others. Fossils also reveal how species have adapted to different environments, providing a **timeline** of life's history and offering a glimpse into the past. Through ongoing research and the discovery of new fossils, we continue to gain insights into the mechanisms of

natural selection, adaptation, and the remarkable diversity of life on Earth.

Part 3: The Power of DNA in Science and Medicine

Genetic Engineering and CRISPR – Editing Life's Code

Genetic engineering refers to the manipulation of an organism's genome using biotechnology, enabling scientists to directly modify an organism's DNA. One of the most groundbreaking advancements in genetic engineering is the development of the **CRISPR-Cas9** system, a precise tool that allows scientists to "edit" genes with remarkable accuracy and efficiency. This technology is revolutionizing fields like medicine, agriculture, and biotechnology, offering the potential to cure genetic diseases, enhance crop production, and even alter the genetic makeup of organisms.

1. What is Genetic Engineering?

Genetic engineering, also known as **genetic modification**, is the process of altering the DNA of an organism to achieve desired traits. This can involve:

Adding new genes to introduce new traits or abilities.

Removing or deactivating specific genes to prevent undesired traits or diseases.

Editing existing genes to correct mutations or enhance natural traits.

Genetic engineering can be applied to a wide range of organisms, including bacteria, plants, animals, and even humans. The techniques used for genetic modification vary, but they all rely on understanding and manipulating the genetic code of life: **DNA**.

2. The Role of CRISPR-Cas9 in Genetic Engineering

CRISPR (Clustered Regularly Interspaced Short Palindromic Repeats) is a naturally occurring defense mechanism found in bacteria. It allows bacteria to recognize and defend against viral infections by "remembering" the DNA of previous invaders. The **CRISPR-Cas9** system is composed of two main components:

CRISPR sequences: Short, repetitive DNA sequences that store the "memories" of past viruses.

Cas9 enzyme: A molecular "scissors" that can cut DNA at specific locations, allowing for the insertion, deletion, or modification of genes.

Scientists harnessed CRISPR-Cas9 for **genetic editing** because of its precision, simplicity, and efficiency. The ability to target and edit specific genes makes it a revolutionary tool in genetic engineering.

3. How CRISPR Works

The CRISPR-Cas9 system works in three key steps:

1. Designing the Guide RNA

Scientists design a custom **guide RNA** (gRNA) that is complementary to the specific DNA sequence they want to edit. This guide RNA acts like a GPS, directing the Cas9 enzyme to the exact location in the genome where the cut should be made.

2. Cutting the DNA

Once the guide RNA binds to its matching DNA sequence, the **Cas9 enzyme** makes a cut in the DNA at that location. This cut creates a **double-strand break** in the DNA, which is crucial for editing.

3. Editing the DNA

After the DNA is cut, the cell's natural repair mechanisms kick in. Scientists can manipulate this repair process in two ways:

Non-homologous end joining (NHEJ): The cell can repair the break in a way that introduces errors, leading to gene disruption. This can be used to deactivate genes or create mutations.

Homology-directed repair (HDR): If a piece of DNA with the desired change is provided, the cell can use this template to repair the break, introducing the desired genetic modification (e.g., inserting a new gene or correcting a mutation).

4. Applications of Genetic Engineering and CRISPR

CRISPR and genetic engineering are being applied across various fields, from medicine to agriculture, offering enormous potential for innovation and solving global challenges.

1. Medicine and Gene Therapy

Gene therapy involves using genetic engineering to treat or prevent diseases by correcting defective genes. CRISPR has shown great promise in medicine, particularly for:

Curing genetic diseases: CRISPR has been used in experimental treatments to correct genetic mutations responsible for diseases like **sickle cell anemia, cystic fibrosis**, and **muscular dystrophy**.

Cancer therapy: Scientists are exploring ways to use CRISPR to edit immune cells, enhancing their ability to target and destroy cancer cells.

HIV: CRISPR is being studied as a way to edit the genes of immune cells, potentially rendering them resistant to the **HIV virus**.

2. Agriculture and Food Production

Genetic engineering has long been used to create crops that are more resistant to pests, diseases, and environmental stress, as well as those that have higher nutritional content. With CRISPR, the potential to create genetically modified crops has become even more powerful:

Disease-resistant crops: CRISPR can be used to create crops resistant to viruses, bacteria, and fungi, reducing the need for harmful pesticides.

Improved nutrition: Genetic modifications can increase the nutritional value of crops, such as enhancing the **vitamin A** content of rice (as seen with **Golden Rice**).

Faster crop development: CRISPR can shorten the time it takes to develop new crop varieties, accelerating the process of agricultural innovation.

3. Biotechnology and Research

In biotechnology, CRISPR has made research more efficient by enabling scientists to edit genes in a wide range of organisms. This allows for the creation of **genetically modified organisms (GMOs)** that are used for studying diseases, developing new treatments, and understanding genetic functions.

Animal models: Scientists use CRISPR to create animal models of human diseases, improving our understanding of diseases like **Alzheimer's, Parkinson's**, and **diabetes**.

Synthetic biology: CRISPR is used to design and engineer microorganisms that can produce pharmaceuticals, biofuels, or other valuable products.

5. Ethical Considerations and Concerns

While CRISPR and genetic engineering hold tremendous potential, they also raise significant ethical concerns, particularly regarding their use in humans and the environment.

1. Germline Editing and Human Cloning

One of the most debated issues is the potential for **germline editing**—modifying the DNA of human embryos or reproductive cells, which would affect future generations. There are concerns about the **unintended consequences**, **designer babies**, and the potential for **genetic inequality**.

Example: In 2018, a Chinese researcher controversially edited the genomes of two embryos to make them resistant to HIV, sparking an ethical outcry and calls for stronger regulation.

2. Environmental Impact

The use of genetic engineering in agriculture, particularly with gene-edited crops, has raised concerns about the impact on ecosystems and biodiversity. There are questions about the potential for **genetically modified organisms (GMOs)** to spread uncontrollably, affecting wild species and disrupting natural ecosystems.

3. Equity and Access

As genetic engineering becomes more powerful, concerns arise about **equity and access**. There is the possibility that only certain parts of the world or certain groups of people will have access to these technologies, leading to further disparities in health and food security.

Genetic engineering and CRISPR have opened a new era in biotechnology, providing unprecedented opportunities to manipulate the genetic code of life. From treating genetic diseases to enhancing crops and advancing scientific research, the applications of these technologies are vast and transformative. However, these advances come with ethical, social, and environmental challenges that must be carefully considered. As we continue to harness the power of **genetic editing**, it is crucial to balance innovation with responsibility to ensure that the benefits of these technologies are realized safely and equitably.

The Human Genome Project – Unlocking Our Genetic Blueprint

The **Human Genome Project (HGP)** was one of the most ambitious and groundbreaking scientific endeavors in history. Launched in 1990 and completed in 2003, this international research initiative aimed to map and sequence the entire **human genome**, unlocking the genetic blueprint of life. The project revolutionized our understanding of genetics, paving the way for advances in **medicine, biotechnology, and evolutionary biology**.

1. What is the Human Genome?

The **human genome** is the complete set of **DNA** in a human cell. It contains all the genetic information necessary for growth, development, and function. The genome consists of:

Approximately 3 billion base pairs of DNA.

About 20,000–25,000 protein-coding genes, which make up only about 1% of the genome.

Non-coding DNA, which was once considered "junk DNA" but is now known to play regulatory and structural roles.

Understanding the genome is essential for deciphering how genes influence health, disease, and human traits.

2. Goals of the Human Genome Project

The Human Genome Project had several key objectives:

1. Mapping the Entire Human Genome

Identify all the genes present in human DNA.

Determine their locations and sequences.

2. Sequencing the DNA

Decode the **3 billion DNA base pairs** to understand their structure and function.

3. Storing and Analyzing Genetic Data

Develop bioinformatics tools to process and interpret massive amounts of genetic information.

4. Identifying Genetic Variations

Discover how genetic differences contribute to diseases and individual traits.

5. Advancing Medicine and Biotechnology

Provide insights into genetic diseases and create new treatments.

3. Key Discoveries and Achievements

The Human Genome Project led to several major breakthroughs:

1. The Complete Human Genome Sequence

The first draft of the human genome was published in 2001.

The final version was completed in **April 2003**, two years ahead of schedule.

2. The Number of Human Genes

Initially, scientists expected humans to have **100,000 genes**. The project revealed that humans have **only around 20,000–25,000 genes**, similar to many other organisms.

3. The Role of Non-Coding DNA

While protein-coding genes make up only **1%** of the genome, the remaining **99%** includes regulatory elements, structural components, and sequences with unknown functions.

4. Genetic Variation and Disease

The project identified **single nucleotide polymorphisms (SNPs)** — small genetic variations that can influence traits and disease susceptibility.

5. Evolutionary Insights

Comparative genomics showed that humans share over **98% of their DNA with chimpanzees** and have genetic similarities with many other species.

4. Impact of the Human Genome Project

The HGP transformed multiple fields, from medicine to biotechnology and forensic science.

1. Medicine and Personalized Healthcare

Enabled the development of **gene-based therapies** for genetic disorders.

Allowed for **personalized medicine**, where treatments can be tailored based on a person's genetic profile.

Improved understanding of diseases like **cancer, Alzheimer's, and diabetes**.

2. Biotechnology and Pharmaceuticals

Led to the creation of new drugs targeting specific genetic pathways.

Improved the production of **gene-based vaccines and therapies**.

3. Ancestry and Evolutionary Biology

Helped trace human **migration patterns** and ancestral origins.

Provided insights into how humans evolved and adapted over time.

4. Forensic Science

Improved **DNA fingerprinting** for solving crimes and identifying individuals.

5. Ethical, Legal, and Social Implications (ELSI)

The Human Genome Project raised important ethical and legal concerns:

1. Genetic Privacy and Discrimination

Who should have access to a person's genetic information?

Could employers or insurance companies discriminate based on genetic data?

In response, the **Genetic Information Nondiscrimination Act (GINA)** was passed in 2008 to protect individuals from genetic discrimination.

2. Ethical Concerns of Genetic Editing

The ability to edit genes raises concerns about **designer babies** and unintended consequences.

3. Ownership of Genetic Data

Debate over whether companies or individuals should own and control genetic information.

6. The Future of Genomics

Since the completion of the Human Genome Project, advances in DNA sequencing technology have made genetic research faster and more affordable.

1. Precision Medicine

Tailoring treatments to individual genetic profiles for diseases like cancer and rare genetic disorders.

2. CRISPR and Gene Editing

The discovery of **CRISPR-Cas9** has made gene editing more precise, raising possibilities for **curing genetic diseases**.

3. The 1000 Genomes Project and Beyond

Researchers are now sequencing the genomes of thousands of individuals to understand genetic diversity and disease susceptibility.

4. Synthetic Biology

Scientists are exploring ways to design synthetic genomes for applications in medicine and industry.

The **Human Genome Project** was a monumental achievement that unlocked the blueprint of life. It revolutionized medicine, biotechnology, and evolutionary biology, setting the stage for future discoveries. While it opened new possibilities for genetic research, it also raised ethical challenges that continue to be debated today. As genomic science advances, the potential for improving human health and understanding life itself is greater than ever.

Genetics and Medicine – From Cancer to Gene Therapy

The field of genetics has transformed modern medicine, offering new insights into **disease causes, diagnosis, and treatment**. Many diseases, from **cancer** to **genetic disorders**, are rooted in changes to **DNA**, and advances in **genomics and gene therapy** are paving the way for more effective, personalized treatments.

1. The Role of Genetics in Medicine

Genetics plays a crucial role in understanding diseases and developing targeted therapies. The discovery of **genes linked to diseases** has led to:

Early diagnosis through genetic testing.

Personalized medicine, where treatments are tailored to a person's genetic profile.

Gene therapy, which seeks to correct faulty genes causing disease.

By studying **genetic mutations**, scientists can better understand conditions such as **cancer, inherited disorders, and rare genetic diseases**.

2. Cancer and Genetics

Cancer is primarily a **genetic disease**, caused by mutations that lead to **uncontrolled cell growth**. These mutations can be:

Inherited (germline mutations) – Passed from parents to children, increasing cancer risk (e.g., **BRCA1/BRCA2** mutations in breast cancer).

Acquired (somatic mutations) – Occur during a person's lifetime due to environmental factors like smoking, radiation, or viral infections.

1. Oncogenes and Tumor Suppressor Genes

Two types of genes play a key role in cancer:

Oncogenes: Mutated genes that cause uncontrolled cell division (e.g., **HER2 in breast cancer**).

Tumor suppressor genes: Normally prevent cancer, but when mutated, they fail to control cell growth (e.g., **TP53 gene**).

2. Genetic Testing for Cancer

Genetic tests can identify mutations that increase cancer risk, allowing for **early detection** and **preventive measures**. For example:

BRCA1 and BRCA2 mutations increase the risk of breast and ovarian cancer.

Lynch syndrome mutations increase the risk of colon cancer.

3. Precision Medicine in Cancer Treatment

Advances in genetics have led to **targeted therapies** that attack specific genetic mutations in cancer cells:

Immunotherapy: Uses the body's immune system to fight cancer (e.g., **CAR-T cell therapy**).

Targeted drugs: Block specific cancer-related proteins (e.g., **Gleevec for leukemia** and **Herceptin for breast cancer**).

3. Genetic Disorders and Gene Therapy

Many diseases are caused by **genetic mutations**, either inherited or spontaneous. These include:

Single-gene disorders: Caused by mutations in a single gene (e.g., **cystic fibrosis, sickle cell anemia, Huntington's disease**).

Polygenic disorders: Involve multiple genes and environmental factors (e.g., **diabetes, heart disease**).

1. Gene Therapy – Fixing Faulty Genes

Gene therapy aims to treat or prevent diseases by modifying an individual's DNA. There are two main approaches:

A. Replacing a Faulty Gene

Example: In **spinal muscular atrophy (SMA)**, gene therapy delivers a functional copy of the **SMN1 gene** to restore muscle function.

B. Editing a Gene Using CRISPR

CRISPR-Cas9 allows scientists to precisely edit **disease-causing mutations**.

Example: Researchers are using CRISPR to correct mutations in **sickle cell anemia** and **muscular dystrophy**.

C. Turning Genes On or Off

Some diseases result from genes being overactive or inactive. Gene therapy can **silence** harmful genes or **activate** beneficial ones.

2. FDA-Approved Gene Therapies

Several gene therapies have been approved for treating genetic diseases:

Luxturna – Treats a rare genetic form of blindness.

Zolgensma – Treats **spinal muscular atrophy**.

Kymriah – A CAR-T cell therapy for leukemia.

4. The Future of Genetics in Medicine

1. Personalized Medicine

With advancements in **genomic sequencing**, doctors can tailor treatments based on a patient's **DNA**, leading to more effective and fewer side-effect-prone therapies.

2. Regenerative Medicine

Using **stem cells** and genetic engineering, researchers aim to regenerate damaged tissues and organs.

3. Ethical Considerations

As genetic medicine advances, ethical concerns arise, such as:

Who should have access to **genetic data**?

Should we allow **germline gene editing** (editing DNA in embryos)?

Genetics is revolutionizing medicine, from **cancer treatment** to **gene therapy**. With continued advancements, the future of medicine will be more **precise, personalized, and effective**, offering hope for previously incurable diseases. However, as we unlock the power of **genetic medicine**, careful ethical considerations will be necessary to ensure responsible and equitable use of these technologies.

Epigenetics – How the Environment Shapes DNA Expression

For a long time, scientists believed that **DNA** was a fixed blueprint, determining everything about an organism. However, research in **epigenetics** has revealed that genes can be turned on or off by environmental factors, lifestyle choices, and even experiences. This field of study shows that **DNA is not destiny** — our environment and behavior can shape how our genes function.

1. What is Epigenetics?

Epigenetics is the study of changes in **gene expression** that do not alter the underlying **DNA sequence**. These changes determine **which genes are active or silent** in different cells at different times.

Every cell in your body contains the **same DNA**, but cells function differently because of epigenetic regulation. For example:

Skin cells and **brain cells** have the same DNA, but they express different genes.

Epigenetics allows a fertilized egg to develop into **specialized cell types** like neurons, muscle cells, or immune cells.

How Does Epigenetics Work?

Epigenetic changes occur through chemical modifications that affect **how genes are read** by the cell. The main mechanisms include:

DNA Methylation

A **methyl group (-CH$_3$)** is added to DNA, usually silencing a gene.

Example: High DNA methylation in **tumor suppressor genes** can lead to cancer by preventing their protective function.

Histone Modification

DNA wraps around **histones** (proteins), and chemical modifications can tighten or loosen the DNA, affecting gene expression.

Example: **Acetylation** of histones often activates genes, while **deacetylation** silences them.

Non-Coding RNA (ncRNA)

Some RNA molecules help regulate which genes are turned on or off.

Example: **MicroRNAs (miRNAs)** can silence genes by preventing protein production.

2. How the Environment Shapes Gene Expression

Epigenetic changes can be triggered by:

1. Diet and Nutrition

Nutrients like **folic acid, vitamin B12, and choline** influence **DNA methylation**.

Example: In **pregnant mothers**, a folate-rich diet can impact the baby's gene expression and development.

Example: Studies on **agouti mice** show that diet can switch a gene on or off, affecting coat color and obesity risk.

2. Stress and Mental Health

Chronic stress alters epigenetic marks in the brain, affecting mood and behavior.

Example: Studies on **rats raised by attentive mothers** show lower stress hormone levels due to **increased gene expression for stress regulation**.

Example: **Post-Traumatic Stress Disorder (PTSD)** has been linked to epigenetic changes that affect brain function.

3. Toxins and Pollution

Exposure to **pesticides, heavy metals, and air pollution** can alter DNA methylation patterns.

Example: **Smoking** causes epigenetic changes in lung cells, increasing the risk of lung cancer.

4. Exercise and Physical Activity

Regular exercise can modify gene expression, improving metabolism and brain health.

Example: Exercise increases the production of **BDNF**, a protein that enhances brain function and protects against neurodegenerative diseases.

5. Parental and Early Life Influences

Experiences during **childhood**, including **parenting style** and **early trauma**, can leave lasting epigenetic marks.

Example: Children who experience **early neglect or abuse** show epigenetic changes in stress-response genes.

3. Epigenetics and Disease

Epigenetic modifications play a key role in **disease development**:

1. Cancer

Abnormal DNA methylation can silence tumor suppressor genes, leading to uncontrolled cell growth.

Example: Many cancers have epigenetic changes that turn off genes responsible for repairing DNA.

2. Neurological Disorders

Alzheimer's disease and **Parkinson's disease** have been linked to epigenetic changes affecting neuron function.

Schizophrenia and depression are influenced by epigenetic regulation of brain genes.

3. Autoimmune Diseases

Epigenetics plays a role in conditions like **lupus, rheumatoid arthritis, and multiple sclerosis**.

Example: DNA methylation changes can affect immune system regulation, leading to an overactive immune response.

4. Aging and Longevity

Over time, epigenetic marks accumulate, contributing to **aging and age-related diseases**.

Some researchers believe that reversing epigenetic changes could **slow down aging**.

4. Can Epigenetic Changes Be Reversed?

Unlike genetic mutations, epigenetic changes are **reversible**. Scientists are exploring ways to modify epigenetic marks for **therapeutic purposes**.

1. Epigenetic Drugs

DNA methylation inhibitors (e.g., azacitidine) are used to treat **blood cancers**.

Histone deacetylase inhibitors (HDAC inhibitors) help reactivate suppressed genes.

2. Lifestyle Interventions

Healthy diet, exercise, and **stress reduction** can positively influence gene expression.

Example: Meditation and mindfulness have been shown to alter epigenetic markers related to stress and inflammation.

3. CRISPR and Epigenome Editing

Scientists are developing tools to **edit epigenetic marks** without altering the DNA sequence.

This could lead to targeted treatments for diseases like **cancer, Alzheimer's, and genetic disorders**.

Epigenetics has revolutionized our understanding of **gene expression and disease**. Unlike genetic mutations, epigenetic changes are **dynamic and reversible**, meaning our lifestyle and environment can actively shape our genetic destiny. As research advances, epigenetics could lead to **new therapies for cancer, mental health disorders, and aging-related diseases**, offering hope for a healthier future.

DNA and Forensics – Solving Crimes with Genetics

DNA has revolutionized forensic science, providing one of the most **powerful tools for criminal investigations**. From identifying suspects to exonerating the innocent, forensic DNA analysis has transformed the justice system.

1. What is Forensic DNA Analysis?

Forensic DNA analysis involves extracting and analyzing **genetic material** from biological evidence found at crime scenes. Every individual (except identical twins) has a **unique DNA profile**, making DNA a powerful tool for **identification and comparison**.

Common Sources of DNA in Forensics

DNA can be recovered from various biological materials, including:

Blood

Saliva (from cigarette butts, envelopes, or drinks)

Hair (with roots attached)

Skin cells (touch DNA from objects like weapons or doorknobs)

Sweat and bodily fluids (semen, urine)

Bone and teeth (useful in cold cases and disasters)

2. How DNA Profiling Works

DNA profiling, also known as **DNA fingerprinting**, involves analyzing specific regions of DNA to create a unique genetic profile. The process includes:

1. Collecting DNA Evidence

Crime scene investigators collect biological samples carefully to avoid contamination.

2. Extracting and Amplifying DNA

Scientists extract DNA from cells.

The **polymerase chain reaction (PCR)** amplifies tiny amounts of DNA, making millions of copies for analysis.

3. Analyzing Short Tandem Repeats (STRs)

Forensic labs examine **short tandem repeats (STRs)** — repeating sequences of DNA that vary between individuals.

The FBI's **Combined DNA Index System (CODIS)** uses **13–20 STR markers** for human identification.

4. Comparing DNA Profiles

Investigators compare crime scene DNA with samples from:

- **Suspects**
- **Victims**
- **DNA databases** (e.g., CODIS)

A match indicates a high probability that the DNA belongs to a particular individual.

3. Applications of DNA in Forensic Investigations

1. Identifying Criminals

DNA evidence can link a suspect to a crime scene, even with tiny samples (touch DNA).

Example: In 1986, forensic DNA was first used in a murder case, leading to the conviction of **Colin Pitchfork** in the UK.

2. Exonerating the Innocent

The **Innocence Project** has used DNA testing to overturn wrongful convictions.

Example: In 1993, **Kirk Bloodsworth**, a death row inmate, was exonerated through DNA evidence.

3. Cold Case Investigations

DNA helps solve decades-old cases by reanalyzing old evidence.

Example: The **Golden State Killer** (Joseph DeAngelo) was arrested in 2018 using forensic genealogy.

4. Identifying Victims in Mass Disasters

DNA is used in **disaster victim identification (DVI)** after plane crashes, terrorist attacks, or natural disasters.

Example: After 9/11, forensic scientists used DNA to identify victims from the World Trade Center debris.

5. Familial DNA Searching

When no direct DNA match is found, investigators search for **partial matches** to identify relatives of the suspect.

Example: In the UK, familial DNA led to the arrest of **Craig Harman** in a manslaughter case.

6. Wildlife and Environmental Forensics

DNA helps combat **wildlife trafficking** by identifying illegal animal products.

Example: DNA analysis has traced **elephant ivory** to specific poaching regions in Africa.

4. Challenges and Limitations of DNA Forensics

1. Contamination and Degradation

DNA evidence can be contaminated if not handled properly.

Environmental factors (heat, moisture) can degrade DNA.

2. Partial or Mixed DNA Samples

Crime scenes often contain DNA from multiple people, making analysis complex.

Advanced techniques like probabilistic genotyping help analyze mixed samples.

3. Ethical and Privacy Concerns

Should law enforcement have access to **private genetic databases** like AncestryDNA?

Some fear **genetic surveillance and misuse of DNA data**.

4. Legal Challenges

DNA evidence must be collected legally and analyzed accurately.

Defense attorneys may challenge **DNA reliability, handling, or lab errors**.

5. The Future of DNA Forensics

1. Rapid DNA Testing

New devices allow DNA analysis in **90 minutes**, useful for real-time crime-solving.

2. Forensic Genealogy

Public DNA databases (e.g., GEDmatch) help solve crimes using distant relatives' DNA.

Example: The **Golden State Killer** case was solved using forensic genealogy.

3. Epigenetic Analysis

Scientists are studying how DNA methylation patterns could reveal a **suspect's age, lifestyle, or recent activities**.

4. DNA Phenotyping

Advanced techniques predict **physical traits** (hair color, eye color, ancestry) from DNA.

Example: The company Parabon Nanolabs creates facial reconstructions from DNA.

DNA forensics has revolutionized criminal investigations, offering unmatched accuracy in identifying suspects and solving crimes. As technology advances, forensic DNA will become even more **powerful, faster, and precise**. However, ethical considerations around **privacy, data security, and legal use** must be carefully addressed to ensure justice is served responsibly.

Part 4: DNA and the Future of Life

DNA and the Future of Life

Advancements in **genetic science** are reshaping our understanding of life itself. From **genome editing** and synthetic biology to personalized medicine and biotechnology, DNA is at the forefront of a scientific revolution that could redefine **health, evolution, and even the future of humanity**.

1. The Power of DNA – Understanding Life's Blueprint

DNA is often called the **"code of life"** because it carries the instructions for building and maintaining all living organisms. The discovery of **the human genome** and breakthroughs in genetic technology have enabled scientists to:

Edit genes to cure diseases.

Engineer crops to resist pests and climate change.

Create synthetic organisms with new capabilities.

Explore the possibility of **designer babies** and genetic enhancements.

As we unlock more of DNA's secrets, we face profound **scientific, ethical, and societal challenges**.

2. Gene Editing – Rewriting Life's Code

1. CRISPR and Precision Gene Editing

CRISPR-Cas9 is a revolutionary tool that allows scientists to **edit DNA with precision**, like a molecular "scissors".

Applications of **CRISPR**:

Curing genetic disorders (e.g., sickle cell disease, cystic fibrosis).

Eradicating inherited diseases from future generations.

Engineering super-resistant crops for food security.

2. Genetic Enhancements – A New Era of Evolution?

Scientists are exploring **human genetic modifications** for:

Stronger immune systems to fight diseases.

Enhanced intelligence or physical abilities.

Slowing down aging by altering cellular processes.

3. DNA in Medicine – The Age of Personalized Healthcare

1. Precision Medicine

Instead of a "one-size-fits-all" approach, **DNA-based medicine** tailors treatments to an individual's genetic makeup.

Example: Cancer therapies like **CAR-T cell therapy** use genetic engineering to program immune cells to fight cancer.

2. Predicting and Preventing Diseases

Genetic testing can identify risks for conditions like **Alzheimer's, heart disease, or breast cancer** (BRCA gene mutations).

Preventative treatments and lifestyle changes based on **genetic risks** can help extend life expectancy.

3. Regenerative Medicine and Cloning

Stem cell therapy and **gene therapy** offer new ways to regenerate damaged tissues.

Cloning technology could be used for organ replacement or even reviving extinct species (e.g., woolly mammoths).

4. Synthetic Biology – Creating Artificial Life

1. Building New Organisms

Scientists are designing **synthetic DNA** to create artificial bacteria and new life forms.

Example: In 2010, researchers at the J. Craig Venter Institute created the first **synthetic cell** by assembling a completely new DNA genome.

2. Bioengineering for Sustainability

Genetically engineered microbes can:

Produce **biofuels** to replace fossil fuels.

Clean up oil spills through **bioremediation**.

Convert waste into useful materials (e.g., biodegradable plastics).

3. The Future of Human Evolution

Could humans **direct their own evolution** by altering DNA?

Some futurists predict that **genetic enhancements, synthetic biology, and AI** could lead to a **post-human future**, where humans are biologically and technologically enhanced.

5. Ethical and Social Implications of DNA Technology

With great power comes great responsibility. As we unlock the potential of DNA, we must consider:

Who controls genetic technologies?

Should we edit human embryos?

Could genetic modifications increase inequality (e.g., "genetic elites")?

How do we ensure genetic privacy and security?

Laws and regulations must evolve alongside DNA technology to balance **innovation with ethical responsibility**.

DNA is not just a molecule; it is the blueprint of life. The ability to **edit, engineer, and understand DNA** is transforming medicine, agriculture, and even the future of humanity. While these advancements hold **immense promise**, they also require careful consideration of the **ethical, social, and biological consequences**.

As we stand on the brink of a **genetic revolution**, the choices we make today will shape **the future of life itself**.

The Ethics of Genetic Manipulation

Genetic manipulation—our ability to **edit, modify, and engineer DNA**—is one of the most powerful scientific advancements in history. It has the potential to cure diseases, improve food security, and even alter human evolution. However, it also raises profound ethical questions about **safety, fairness, consent, and the very nature of life**. Should we **intervene in the genetic code of future generations**? Could genetic modification **deepen social inequalities**? Are we **"playing God"**?

This chapter explores the **moral, social, and philosophical dilemmas** surrounding genetic manipulation.

1. The Benefits of Genetic Manipulation

1.1 Eradicating Genetic Diseases

Technologies like **CRISPR-Cas9** allow scientists to edit **disease-causing genes** in embryos, preventing inherited conditions such as:

Cystic fibrosis

Huntington's disease

Sickle cell anemia

Muscular dystrophy

Ethical Debate: Is it our moral duty to prevent suffering through gene editing, or should we accept natural genetic variation?

1.2 Advancing Medicine and Longevity

Gene therapies are being developed to treat **cancer, Alzheimer's, and heart disease**.

Anti-aging research explores ways to slow cellular aging, potentially extending the human lifespan.

Could we create **a future where people live longer, healthier lives** through genetic enhancements?

1.3 Improving Agriculture and Sustainability

Genetically modified (GM) crops can:

Resist pests and diseases

Withstand extreme weather (drought-resistant rice, flood-resistant wheat)

Reduce pesticide use and environmental impact

Lab-grown meat and bioengineered food could address world hunger and climate change.

2. Ethical Concerns of Genetic Manipulation

2.1 Designer Babies – Where Do We Draw the Line?

Therapeutic vs. Enhancement Debate: Should we only fix genetic disorders, or should parents be allowed to **choose traits like intelligence, height, or eye color**?

The Risk of Genetic Inequality: Could genetic enhancements create a **"genetic elite"**, where only the wealthy have access to intelligence or physical enhancements?

The Right to an Unaltered Genome: Should children have a say in their genetic modifications, or is it ethical for parents to make these choices?

2.2 Unintended Consequences – The Risks of Editing DNA

Off-target mutations: CRISPR and other gene-editing tools could accidentally introduce harmful mutations.

Long-term effects are unknown: Altering DNA may have consequences that only appear in future generations.

Loss of genetic diversity: Widespread genetic modification could make humanity vulnerable to new diseases or environmental changes.

2.3 Playing God – Should Humans Control Evolution?

Religious and philosophical objections argue that **life's natural course should not be altered**.

Others believe that **if we have the power to cure suffering, we have an obligation to use it**.

Bioethics dilemma: Where do we draw the line between necessary medical intervention and unnecessary interference with nature?

3. Societal Implications of Genetic Engineering

3.1 Genetic Discrimination and Privacy

Could employers, insurance companies, or governments **discriminate** based on a person's genetic risks?

Genetic surveillance: Some fear that national DNA databases could be misused for tracking or control.

3.2 Genetic Engineering and Social Inequality

If **gene therapy and enhancements** are expensive, will only the rich benefit, widening the gap between wealthy and poor?

Could we create a future where **genetically enhanced individuals** have an unfair advantage in education, sports, and the job market?

3.3 The Threat of Bioweapons and Misuse

Could gene-editing technology be used to create **biological weapons** or **designer viruses**?

International **biosecurity laws** must be enforced to prevent misuse.

4. Regulating Genetic Manipulation – Finding Ethical Boundaries

4.1 The Need for Global Standards

Some countries **allow genetic modification**, while others ban it.

Should there be **a global agreement** on how far genetic engineering should go?

4.2 Balancing Innovation and Ethics

Strict regulations can slow scientific progress.

Lack of regulation can lead to dangerous experiments or unethical practices.

4.3 Case Study: The CRISPR Babies Controversy

In 2018, Chinese scientist **He Jiankui** edited the genes of twin babies to make them resistant to HIV.

The experiment sparked global outrage, and He was sentenced to prison.

Lesson: Gene editing must be **carefully regulated** to prevent unethical experiments.

5. The Future of Genetic Manipulation – Ethical Progress or Dangerous Path?

The **potential of genetic manipulation** is immense, but so are the **ethical risks**.

The challenge is to **use this technology responsibly**, ensuring it benefits all of humanity without creating new forms of inequality or harm.

Public debate, strict regulations, and ethical oversight will be essential as we move forward in the genetic revolution.

Cloning and Synthetic Biology – Creating Life in the Lab

Scientific advancements in cloning and synthetic biology are transforming our ability to manipulate life at its most fundamental level. From cloning entire organisms to designing artificial genes, these technologies offer groundbreaking possibilities for medicine, agriculture, and even species conservation. However, they also raise profound ethical and philosophical questions: Should we create life in the lab? What are the risks of engineering living organisms? Could synthetic life reshape evolution?

This chapter explores the science, potential benefits, and ethical dilemmas of cloning and synthetic biology.

1. Cloning – Copying Life

1.1 What is Cloning?

Cloning is the process of creating a genetically identical copy of an organism. There are three main types of cloning:

Reproductive cloning – Producing a whole organism with identical DNA (e.g., Dolly the sheep).

Therapeutic cloning – Creating cloned cells for medical treatments and organ regeneration.

Gene cloning – Copying specific genes for research or medicine.

1.2 The Science of Cloning – How It Works

The most well-known method of cloning is somatic cell nuclear transfer (SCNT):

A nucleus is extracted from a donor cell.

This nucleus is inserted into an egg cell with its original DNA removed.

The egg is stimulated to divide, forming an embryo with the donor's genetic material.

The embryo is implanted into a surrogate mother (for reproductive cloning) or used for stem cell research (for therapeutic cloning).

1.3 Famous Cloning Experiments

Dolly the Sheep (1996) – The first mammal cloned from an adult cell, proving that cloning was possible.

Cloned pets and livestock – Scientists have cloned cats, dogs, cows, and even endangered species.

Cloning extinct species ("de-extinction") – Efforts are underway to revive the woolly mammoth using elephant DNA.

1.4 Potential Benefits of Cloning

Medical advancements – Cloning could create personalized stem cells to treat diseases like Parkinson's or diabetes.

Preserving endangered species – Cloning could help prevent extinction.

Livestock cloning – Producing disease-resistant, high-yield animals for food production.

1.5 Ethical and Scientific Concerns

Low success rates and health risks – Many cloned embryos fail, and cloned animals often suffer health issues.

Human cloning debate – Cloning humans is widely considered unethical and illegal in most countries.

Loss of genetic diversity – Cloning could reduce natural variation, making populations more vulnerable to disease.

2. Synthetic Biology – Engineering New Life Forms

2.1 What is Synthetic Biology?

Synthetic biology is the design and creation of new biological systems or the reprogramming of existing organisms for useful purposes. Unlike cloning, which copies existing life, synthetic biology builds life from scratch using engineered DNA.

2.2 How Synthetic Biology Works

Scientists use genetic engineering techniques to:

Create synthetic DNA sequences that do not exist in nature.

Insert these genes into bacteria, yeast, or other organisms to give them new functions.

Design biological circuits, similar to computer circuits, to control cellular behavior.

2.3 Key Advances in Synthetic Biology

First synthetic cell (2010) – Scientists at the J. Craig Venter Institute created the first synthetic life form by assembling an entirely new genome.

Artificial genetic code – Researchers have expanded the genetic alphabet beyond the natural A, T, C, and G base pairs, creating new possibilities for synthetic life.

Self-replicating synthetic organisms – Scientists are developing cells that can reproduce and evolve on their own.

2.4 Applications of Synthetic Biology

Medicine and Healthcare

Custom-designed microbes to produce insulin, vaccines, or cancer-fighting drugs.

Gene therapy using synthetic DNA to treat genetic disorders.

Environmental Solutions

Bioengineered bacteria to clean up oil spills or absorb carbon dioxide.

Synthetic algae to produce biofuels and reduce reliance on fossil fuels.

Food Production and Agriculture

Lab-grown meat made from cultured cells, reducing the need for factory farming.

Genetically modified crops engineered for higher yields and pest resistance.

3. Ethical and Safety Concerns of Cloning and Synthetic Biology

3.1 Playing God – Should We Create Life?

Some argue that engineering life forms is an overreach of human power.

Others believe synthetic biology can solve global challenges in health, food, and the environment.

3.2 Potential Risks – What Could Go Wrong?

Unintended consequences – Synthetic organisms might mutate or interact unpredictably with ecosystems.

Biosecurity threats – Could synthetic biology be used to create bioweapons or harmful pathogens?

3.3 The Ethics of Human Genetic Engineering

Should we use synthetic biology to enhance human traits (intelligence, strength, longevity)?

Could genetic modifications lead to inequality, with "enhanced" vs. "natural" humans?

3.4 Regulatory Challenges

Different countries have varying laws on genetic engineering.

4. The Future of Cloning and Synthetic Biology

Could we create entirely new life forms with unique biological systems?

Will synthetic humans be possible? – Some predict we may one day design customized human cells for medical use.

How will cloning and synthetic biology change evolution? – If humans direct the genetic future, will natural evolution become obsolete?

DNA and Artificial Intelligence – The Intersection of Biology and Tech

The fusion of DNA science and artificial intelligence (AI) is revolutionizing biology, medicine, and genetics. AI is accelerating discoveries in genomics, helping scientists decode the human genome, predict diseases, design drugs, and even create synthetic life. But as AI-powered genetics advances, ethical concerns arise: Who owns genetic data? Could AI-designed DNA be dangerous? Will AI help us cure diseases or engineer the next phase of evolution?

This chapter explores how AI is reshaping DNA research, medicine, and the future of life itself.

1. AI and Genomics – Decoding Life's Code Faster Than Ever

1.1 How AI is Transforming DNA Research

AI is revolutionizing genetics by:

Analyzing massive genomic datasets faster than humans ever could.

Identifying patterns and mutations linked to diseases.

Predicting protein structures and functions, accelerating drug discovery.

Designing synthetic genes for medicine, agriculture, and biotech.

1.2 The Human Genome and AI – A Perfect Match

The Human Genome Project took 13 years to map human DNA (1990-2003).

Today, AI-powered sequencing can decode an entire genome in hours.

AI helps researchers compare millions of genomes to find hidden links between genes and diseases.

1.3 AI and CRISPR – Precision Gene Editing

AI is improving CRISPR-Cas9 gene editing by predicting off-target effects.

AI helps scientists design safer, more precise genetic modifications.

Future AI-driven CRISPR could cure genetic diseases like sickle cell anemia and Huntington's disease.

2. AI in Medicine – Personalized Genetics and Disease Prediction

2.1 AI-Powered Genetic Testing

Companies like 23andMe and AncestryDNA use AI to analyze DNA for ancestry and health insights.

AI-driven polygenic risk scores can predict a person's risk of diseases like:

Cancer

Alzheimer's

Heart disease

Diabetes

2.2 AI and Cancer Genomics

AI is helping scientists identify cancer-causing mutations in DNA.

AI-powered diagnostics can detect early-stage tumors from genetic data.

AI tailors personalized cancer treatments based on a patient's unique DNA.

2.3 AI and Drug Discovery

Traditional drug discovery takes 10+ years and billions of dollars.

AI simulates how drugs interact with genes, designing new medicines faster.

AI-designed drugs are already in clinical trials for conditions like ALS and Parkinson's.

3. AI and Synthetic Biology – Creating Life with Algorithms

3.1 AI-Designed DNA and Synthetic Organisms

AI can generate synthetic DNA sequences to create new biological systems.

AI-designed bacteria could be used for:

> Cleaning up pollution (bioremediation)
>
> Producing biofuels and sustainable materials
>
> Synthesizing new medicines

3.2 AI and the Creation of Artificial Cells

Scientists are using AI to design synthetic cells from scratch.

AI models predict how synthetic DNA will behave, reducing trial and error.

Future AI-created organisms could perform tasks like manufacturing or self-repair.

4. The Ethical Dilemmas of AI in Genetics

4.1 Who Owns Genetic Data?

AI companies collect massive amounts of genetic data from individuals.

Privacy concerns: Could AI-designed health predictions be misused?

Genetic discrimination: Could insurance companies or employers use DNA data unfairly?

4.2 AI and Genetic Engineering – Are We Playing God?

Should AI be allowed to design human DNA?

Could AI create genetic enhancements that widen inequality?

If AI can predict genetic diseases before birth, could it lead to eugenics concerns?

4.3 AI-Designed DNA – Risks of Synthetic Life

Could AI create unintended mutations or dangerous biological weapons?

Regulation challenges: Who decides which AI-driven genetic modifications are ethical?

5. The Future – AI, DNA, and the Next Evolutionary Leap

AI is merging with biology, creating a new era of bioengineering.

AI-powered DNA research could lead to cures for genetic diseases, longer lifespans, and even synthetic life forms.

The ultimate question: Will AI help us master genetics responsibly, or are we heading toward a future of unintended consequences?

The Future of Human Evolution – Where Are We Headed?

Humanity has been shaped by **natural selection**, but the forces driving our evolution are changing. **Genetic engineering, artificial intelligence, climate change, and space colonization** could transform the future of our species. Are we still evolving? Will we become a **genetically enhanced species**? Could we one day merge with machines?

This chapter explores the possible futures of **human evolution**, from natural changes to bioengineering and transhumanism.

1. Are Humans Still Evolving?

1.1 The Evidence for Ongoing Evolution

Scientists have found **genetic changes** in modern humans, showing that evolution is still happening:

> **Lactose tolerance** evolved in some populations in the past 10,000 years.
>
> **High-altitude adaptation** in Tibetans and Andean peoples.
>
> **Disease resistance** mutations, such as protection against malaria.

Evolutionary pressures have shifted: Instead of survival of the fittest, modern humans experience:

> **Medical advances** that allow people to survive genetic conditions.
>
> **Cultural and technological selection**, where intelligence and creativity matter more than physical strength.
>
> **Social and environmental changes**, influencing genetic trends.

1.2 Will Evolution Slow Down?

Some scientists believe **modern medicine and globalization** have reduced natural selection.

However, **new evolutionary pressures** — like climate change, pollution, and technology — may shape future adaptations.

2. The Next Phase – Technology and Bioengineering

2.1 Genetic Engineering and CRISPR

CRISPR and gene editing allow us to **modify human DNA**, eliminating genetic diseases.

Future possibilities:

> **Disease resistance** – Editing genes to prevent cancer, Alzheimer's, and heart disease.
>
> **Enhanced intelligence or strength** – Could future humans be designed for superior abilities?
>
> **Longevity genes** – Could aging be slowed or even reversed?

2.2 Artificial Intelligence and Human-Machine Integration

Brain-computer interfaces (BCIs) like Elon Musk's **Neuralink** could connect human minds with AI.

Possible outcomes:

> **Superintelligence** – Enhanced cognitive abilities.
>
> **Memory upload/download** – Storing memories like data.
>
> **Mind-melding** – Direct brain-to-brain communication.

2.3 The Rise of Transhumanism

Transhumanists believe humans should **use technology to enhance evolution**.

Possible transformations:

Cyborgs – Integrating robotic limbs, artificial organs, and enhanced senses.

Digital consciousness – Uploading the human mind to computers.

Synthetic biology – Designing new biological traits from scratch.

3. Environmental and Cosmic Evolution

3.1 Climate Change and Adaptation

Rising temperatures and pollution could drive **physical and genetic changes**:

Darker skin tones for UV protection.

Larger lung capacities for polluted environments.

Smaller bodies for better heat regulation.

3.2 Evolution Beyond Earth – Becoming a Spacefaring Species

Colonizing Mars and beyond could lead to a **new branch of human evolution**.

Possible adaptations for space living:

Denser bones and muscles to withstand low gravity.

Radiation resistance to survive cosmic rays.

Oxygen-efficient metabolisms for thin atmospheres.

3.3 Speciation – Could New Human Species Emerge?

If humans settle different planets, could we **diverge into multiple species**?

Over thousands of years, **Martian humans** could look and function differently from Earth-based humans.

4. The Ethical and Philosophical Questions

4.1 Should We Direct Human Evolution?

If we **control genetics and AI**, do we stop being "natural" humans?

Could **designer babies** and genetic enhancements create social inequality?

4.2 The Future of Identity – What Makes Us Human?

If humans merge with AI, will we still be **human** or something else?

Could we become a **post-human civilization**—a species beyond biological constraints?

5. The Future – A New Chapter in Human Evolution

Whether through **natural selection, genetic engineering, or technology**, human evolution is far from over.

The future could lead to **longer lifespans, new abilities, and even new species**.

The ultimate question: **Are we ready for the next step in evolution, or will we lose what makes us human?**

Conclusion: The Code of Life and Our Place in the Universe

The study of **DNA, evolution, and genetics** has revealed profound insights into the **nature of life**, from its origins to its future. We've unraveled the mysteries of how life works, how it changes, and how we can potentially **direct its future**. The **genetic code** is the blueprint of life, intricately designed and passed down through generations, yet it's not a static text—it is subject to change, adaptation, and sometimes, even manipulation.

As we stand at the crossroads of biological, technological, and ethical frontiers, we face the **ultimate questions about our place in the universe**. What does it mean to be human in a world where **genetic engineering**, **artificial intelligence**, and **space exploration** offer unprecedented possibilities for **self-transformation**? Will we harness our newfound abilities to **enhance life**, or will we risk unraveling the fabric of **evolution itself**?

1. A New Era of Possibilities

Humanity stands at the threshold of a transformative age, where innovation, technology, and shifting mindsets are redefining what is possible. From advancements in artificial intelligence to breakthroughs in medicine, space exploration, and sustainable living, we are witnessing an era that is opening doors once thought to be locked by the constraints of the past.

Technology: The Driving Force of Change

The rapid evolution of technology has reshaped the way we live, work, and interact. Artificial intelligence is revolutionizing industries, automating processes, and enhancing efficiency in ways previously unimaginable. Quantum computing promises to solve complex problems at speeds far beyond today's capabilities. Meanwhile, biotechnology is paving the way for personalized medicine, potentially curing diseases that have plagued humanity for centuries.

Expanding the Frontiers of Space Exploration

Space travel is no longer just the domain of government agencies. Private enterprises are pushing the boundaries, with ambitious plans to colonize Mars, mine asteroids, and create sustainable space habitats. This new era of exploration is not just about curiosity but about securing the future of our species beyond Earth.

Sustainability and a Greener Future

The urgency of climate change has propelled industries and individuals toward sustainable solutions. Renewable energy, electric transportation, and circular economies are no longer niche concepts but mainstream movements. The transition toward a greener world is not just an environmental necessity but a testament to human ingenuity and resilience.

A Shift in Human Consciousness

Beyond technology and science, we are witnessing a shift in human consciousness. There is a growing emphasis on inclusivity, mental well-being, and purpose-driven innovation. Societies are recognizing the value of diversity, collaboration, and ethical progress, fostering a more connected and compassionate world.

Embracing the Future with Optimism

The challenges ahead are undeniable, but the possibilities are limitless. As we navigate this new era, the key lies in embracing change with curiosity, responsibility, and a commitment to progress. The future is not just something we step into—it is something we shape.

We are not just entering a new era; we are creating it.

2. The Intersection of Nature and Technology

As humanity advances, the relationship between nature and technology has evolved from one of conflict to collaboration. Today, we stand at a crucial juncture where innovation and ecological awareness are merging to create a more sustainable, efficient, and harmonious future. The intersection of nature and technology is reshaping industries, conservation efforts, and the way we interact with the environment.

The Evolution of Human-Nature Interaction

For centuries, technological progress often came at the cost of environmental degradation. The Industrial Revolution led to deforestation, pollution, and habitat destruction. However, in recent decades, a shift has occurred — technology is no longer just exploiting nature but working alongside it. Scientists, engineers, and environmentalists are now leveraging advanced technologies to preserve ecosystems, combat climate change, and create nature-inspired solutions.

Biomimicry: Learning from Nature's Designs

Biomimicry is one of the most fascinating ways technology and nature intersect. It involves studying nature's time-tested solutions to solve human challenges. Some groundbreaking examples include:

Velcro: Inspired by burrs that stick to animal fur, Velcro was developed to mimic the tiny hooks found in nature.

Self-Cleaning Surfaces: The lotus leaf's ability to repel water and dirt has led to the development of self-cleaning paints and materials.

Efficient Wind Turbines: Engineers designed more efficient wind turbines by studying the aerodynamic shape of humpback whale fins.

By observing and replicating nature's designs, scientists are improving efficiency, sustainability, and resilience in various fields.

Sustainable Energy: Harnessing Nature's Power

The world is moving away from fossil fuels and embracing renewable energy sources inspired by nature. Innovations in this sector include:

Solar Energy: Photovoltaic cells mimic the way plants absorb sunlight for photosynthesis. Solar panel efficiency continues to improve with advancements in nanotechnology.

Wind Energy: Wind farms, designed to replicate the flow patterns of bird flight and school fish movement, increase energy efficiency.

Hydropower and Ocean Energy: Tidal and wave energy projects are utilizing nature's water currents to generate electricity, offering a sustainable alternative to traditional power sources.

These technologies highlight how we can harness nature's forces without disrupting ecosystems.

Smart Agriculture: A Tech-Driven Green Revolution

Agriculture, one of humanity's oldest industries, is now being revolutionized by technology to increase sustainability and reduce waste. Key developments include:

Vertical Farming: Using hydroponics and aeroponics, vertical farms reduce land and water usage while producing crops year-round in controlled environments.

Precision Farming: Drones, sensors, and AI-driven analytics help farmers optimize irrigation, monitor crop health, and reduce pesticide use, leading to higher yields with minimal environmental impact.

Lab-Grown and Alternative Proteins: Cellular agriculture is producing lab-grown meat, reducing the need for livestock farming, which is a major contributor to greenhouse gas emissions.

By integrating technology with agriculture, we can meet growing food demands while minimizing environmental harm.

Conservation Technology: Protecting Nature with Innovation

Conservation efforts have been greatly enhanced through technological advancements. Key examples include:

AI and Machine Learning: Scientists use AI to analyze satellite images and monitor deforestation, track endangered species, and detect illegal poaching activities.

DNA and Genetic Technologies: Genetic tools like CRISPR and environmental DNA (eDNA) are being used to protect biodiversity, revive extinct species, and understand ecosystems at a molecular level.

Bioacoustics and IoT Sensors: Networks of sensors and AI-driven audio analysis help researchers study animal communication, detect habitat changes, and protect species in real time.

These technologies are giving conservationists powerful tools to preserve wildlife and ecosystems more effectively than ever before.

Eco-Friendly Urban Design: The Cities of the Future

The intersection of nature and technology is also transforming how cities are designed. Urban planners are integrating nature into the built environment through:

Green Architecture: Buildings covered in living plants, such as vertical gardens and green roofs, improve air quality, reduce heat, and enhance biodiversity in urban areas.

Smart Cities: AI-driven traffic management, energy-efficient lighting, and IoT-powered waste management systems make cities more sustainable.

Sustainable Infrastructure: Materials like self-healing concrete (inspired by bacteria) and algae-powered streetlights are reducing urban environmental footprints.

With these innovations, cities are evolving into more sustainable, livable, and eco-conscious spaces.

Artificial Intelligence and Big Data for Environmental Solutions

AI and big data are playing a crucial role in environmental research and action. Some major applications include:

Climate Modeling: AI-driven simulations predict climate change patterns and help policymakers create data-driven strategies.

Wildfire Detection: AI-powered satellite monitoring systems detect wildfires before they spread, enabling faster response times.

Pollution Control: Machine learning algorithms analyze air and water quality data to identify pollution sources and suggest remediation methods.

By analyzing vast amounts of environmental data, AI is helping scientists and decision-makers create more effective sustainability strategies.

1. Climate Change Monitoring and Mitigation

AI and Big Data are transforming climate science by providing accurate predictions and insights into global warming patterns. Key applications include:

Climate Modeling: AI-powered simulations predict future climate scenarios, helping policymakers make informed decisions.

Carbon Footprint Reduction: AI optimizes energy use in industries, reducing greenhouse gas emissions. Smart grids, for example, enhance energy efficiency.

Early Warning Systems: AI analyzes weather patterns and historical data to predict extreme weather events such as hurricanes, floods, and wildfires, giving communities time to prepare.

2. Wildlife Conservation and Biodiversity Protection

Protecting biodiversity is critical for maintaining ecological balance. AI and Big Data support conservation efforts through:

AI-Powered Monitoring: Drones and satellite imagery, combined with AI, track wildlife populations, detect deforestation, and monitor illegal activities like poaching.

Acoustic Sensors for Species Detection: AI analyzes sounds in forests and oceans to identify species and track their movements, helping to protect endangered animals.

Predictive Analytics: Machine learning models forecast habitat loss and suggest conservation strategies.

3. Sustainable Agriculture and Food Security

Agriculture is both a driver of environmental change and a sector that can benefit from AI-driven sustainability. Key innovations include:

Precision Farming: AI analyzes soil conditions, weather patterns, and crop health to optimize irrigation and fertilizer use, reducing waste and improving yields.

Pest and Disease Detection: AI-powered image recognition identifies pests and plant diseases early, reducing the need for harmful pesticides.

Supply Chain Optimization: Big Data ensures food production and distribution are more efficient, reducing food waste.

4. Smart Cities and Sustainable Urban Planning

AI and Big Data are shaping the cities of the future by making them greener and more efficient. Examples include:

Traffic Optimization: AI-powered traffic systems reduce congestion and emissions by adjusting traffic signals in real time.

Waste Management: Smart sensors monitor waste levels, enabling efficient collection and recycling efforts.

Green Building Design: AI-driven simulations model energy-efficient building designs to minimize environmental impact.

5. Water and Air Quality Management

Ensuring clean water and air is crucial for human health and the environment. AI helps by:

Real-Time Pollution Tracking: AI-powered sensors detect air and water pollution, allowing for rapid responses.

Smart Water Management: AI predicts water demand, detects leaks, and optimizes irrigation systems, conserving water resources.

Ocean Cleanup Efforts: AI-driven robotic systems collect plastic waste from oceans, helping to combat marine pollution.

Ethical Considerations and Challenges

While AI and Big Data offer immense environmental benefits, their implementation must be guided by ethical principles. Key concerns include:

Data Privacy: Large-scale environmental monitoring may raise concerns about surveillance and data security.

Algorithm Bias: AI models must be trained on diverse datasets to ensure fair and accurate environmental decision-making.

Energy Consumption of AI: Training AI models requires significant computational power, potentially increasing carbon emissions if not managed sustainably.

Access and Equity: Developing countries may have limited access to AI and Big Data technologies, creating disparities in environmental protection efforts.

A Smarter, Greener Future

Artificial Intelligence and Big Data are transforming the way we tackle environmental challenges, offering innovative solutions for climate change, conservation, and sustainability. However, ethical considerations must be addressed to ensure these technologies benefit all of humanity without causing unintended harm.

As AI continues to evolve, its potential for environmental protection is limitless. By embracing responsible innovation, we can harness the power of AI and Big Data to create a more sustainable future for our planet.

The Ethical Considerations of Merging Nature and Technology

As we integrate technology with nature, ethical considerations must be addressed. Key concerns include:

Biotechnology Risks: While genetic engineering and synthetic biology have potential benefits, they also raise concerns about unintended consequences for ecosystems.

Data Privacy in Conservation: AI and satellite monitoring raise ethical questions about privacy, land ownership, and the rights of indigenous communities.

Sustainable Development Balance: Striking the right balance between technological advancement and environmental preservation is crucial to prevent over-exploitation of natural resources.

Addressing these ethical challenges requires responsible innovation and collaboration between scientists, policymakers, and communities.

The fusion of nature and technology takes many forms, including:

Genetic Engineering: CRISPR and other gene-editing tools allow scientists to modify DNA, potentially eradicating diseases or creating new life forms.

Artificial Intelligence in Conservation: AI is used to monitor wildlife, prevent poaching, and optimize farming to reduce environmental damage.

Bioengineering and Synthetic Life: Scientists are creating lab-grown organs, engineered bacteria, and even artificial ecosystems.

Cyborg and Human Enhancement Technologies: Neural implants, prosthetics, and brain-computer interfaces are bridging the gap between humans and machines.

Each of these innovations holds the potential for enormous benefits but also comes with ethical dilemmas.

Key Ethical Concerns

1. Playing God: Should We Modify Life?

Altering the genetic structure of organisms raises fundamental questions about the limits of human intervention. Should we modify crops to be resistant to disease? Should we bring back extinct species? While such actions could benefit humanity, they also disrupt natural evolution and may have unintended consequences.

2. Ecological Disruptions and Unintended Consequences

Genetically Modified Organisms (GMOs): While GMOs can increase food production, they may also impact biodiversity, creating imbalances in ecosystems.

Bioengineered Species: Introducing lab-created organisms into the wild could have unpredictable effects on natural ecosystems.

Geoengineering: Large-scale efforts to modify the climate, such as artificial clouds to cool the Earth, carry the risk of unforeseen side effects.

3. The Ethical Treatment of Enhanced and Artificial Life

If we create sentient AI or bioengineered life forms, what rights should they have?

Should AI-powered systems that mimic nature (such as robotic bees for pollination) be considered part of the ecosystem?

If we merge human brains with machines, does that redefine what it means to be human?

These questions challenge traditional ethical frameworks and force us to reconsider our moral responsibilities.

4. Inequality and Accessibility

Who gets access to these advancements? If human genetic enhancement, cybernetic implants, or life-extension technologies become widely available, will they be accessible to all, or only to the wealthy? The risk of creating a technological divide between the "enhanced" and the "natural" could deepen social inequalities.

5. Corporate and Government Control

The merging of technology with nature often falls under the control of corporations and governments, raising concerns about:

Biopiracy: The exploitation of natural genetic resources by private companies without fair compensation to indigenous communities.

Privacy and Surveillance: AI-driven environmental monitoring could be misused for mass surveillance.

Weaponization of Biotechnology: Gene editing and synthetic biology could be used for harmful purposes, such as biological warfare.

Finding a Balance: Ethical Guidelines for the Future

To ensure responsible integration of technology and nature, we must establish ethical frameworks that prioritize:

Sustainability: Ensuring that technological advancements support, rather than harm, ecosystems.

Transparency and Regulation: Creating clear policies for genetic modification, AI use, and synthetic biology.

Fair Access: Preventing technological advancements from becoming tools of inequality.

Respect for Life: Recognizing the value of both natural and artificial life forms.

Shaping a Responsible Future

The merging of nature and technology offers incredible possibilities for improving human life and the environment. However, without careful ethical consideration, it could also lead to unintended harm, inequality, and ecological disruption. The key challenge lies in harnessing these advancements responsibly—respecting both the natural world and the moral boundaries of technological progress.

As we continue to blur the lines between nature and innovation, we must ask ourselves: Are we shaping the future wisely, or are we blindly altering the foundations of life? The ethical choices we make today will determine the legacy we leave for future generations.

The Future: A Symbiotic Relationship Between Nature and Technology

The future of technology is not about conquering nature but coexisting with it. Some emerging trends that will further strengthen this relationship include:

Regenerative Design: Creating technology that restores and improves ecosystems rather than merely sustaining them.

Biodigital Integration: The merging of biological and digital technologies to create self-sustaining and adaptive solutions for both human and environmental needs.

Decentralized, Nature-Inspired Systems: Moving away from centralized industries to community-driven, sustainable models based on natural ecosystems.

As we continue to innovate, the key to success will be respecting and learning from the natural world rather than working against it.

Conclusion: A Future Where Nature and Technology Thrive Together

The intersection of nature and technology is not just about sustainability—it's about reimagining how we interact with the world around us. By drawing inspiration from nature, harnessing its forces responsibly, and using technology to protect and restore ecosystems, we can build a future where both humanity and the environment thrive.

The future is not about choosing between nature and technology but about finding ways for them to work in harmony. Through smart innovation, ethical responsibility, and a deep respect for the natural world, we are entering an era where nature and technology no longer stand at odds but move forward as partners in progress.

3. The Universe Awaits

Since the dawn of time, humanity has looked to the stars with wonder, curiosity, and a longing to explore the unknown. The vast expanse of the universe—filled with galaxies, stars, planets, and mysteries yet to be uncovered—beckons us forward. As we stand on the brink of an interstellar future, the universe awaits, inviting us to push the boundaries of science, technology, and human potential.

A Cosmic Perspective: Our Place in the Universe

Earth is but a tiny speck in the grand cosmic ocean, orbiting an average star in a galaxy of hundreds of billions of stars, within a universe that may be infinite. The scale of the cosmos is overwhelming, yet it is also a reminder of the boundless possibilities that lie ahead. Every star may have planets, and among them, worlds that could host life—or become the future homes of humankind.

The Age of Space Exploration: The Next Frontier

We have already taken our first steps into the cosmos. The Moon landings of the 20th century, the Mars rovers, and the ongoing efforts of space agencies and private enterprises are paving the way for a new era of exploration. Key milestones on the horizon include:

Human Missions to Mars: NASA, SpaceX, and other organizations are planning crewed missions to the Red Planet, aiming to establish a sustainable presence beyond Earth.

Lunar Colonization: With the Artemis program and private ventures, a permanent human presence on the Moon is within reach, serving as a stepping stone for deeper space travel.

Interstellar Travel: Concepts like the Breakthrough Starshot initiative, nuclear fusion propulsion, and even theoretical warp drives suggest that travel beyond our solar system could one day become reality.

The universe is not just something to observe—it is something to explore.

The Search for Extraterrestrial Life: Are We Alone?

One of the greatest questions of all time is whether life exists beyond Earth. With billions of potentially habitable planets in our galaxy alone, the probability suggests that we are not alone. Scientists are searching for answers through:

Exoplanet Research: The James Webb Space Telescope and other missions are scanning distant planets for signs of atmospheres and conditions suitable for life.

SETI (Search for Extraterrestrial Intelligence): Radio telescopes listen for potential signals from intelligent civilizations.

Mars and Europa Missions: Spacecraft are searching for microbial life in subsurface oceans and ancient riverbeds.

Finding life beyond Earth would be one of the most profound discoveries in human history, reshaping our understanding of the cosmos and our place in it.

The Future of Humanity in Space

If humanity is to survive in the long term, we must become a multi-planetary species. The threats of asteroid impacts, climate change, and other existential risks make space colonization not just an adventure but a necessity. Future developments could include:

Space Habitats: Giant orbiting space stations, like those envisioned by physicist Gerard K. O'Neill, could sustain human civilizations beyond Earth.

Terraforming: Transforming Mars or other celestial bodies into habitable worlds may one day be possible through advanced technology.

Asteroid Mining: Harvesting resources from asteroids could provide limitless materials for building and energy needs.

The future is not limited to Earth. The universe offers infinite opportunities for expansion and evolution.

A Call to Explore: The Universe is Ours to Discover

The universe awaits—filled with mysteries to unravel, planets to explore, and new frontiers to conquer. Our journey into the cosmos is just beginning, and the challenges ahead are vast. Yet, just as our ancestors once braved the open seas, so too must we reach for the stars.

4. The Ethics of the Code of Life

The ability to manipulate the very code of life—DNA—has ushered humanity into a new era of scientific progress. With advancements in genetic engineering, synthetic biology, and biotechnology, we now have the power to alter genes, cure diseases, and even design living organisms. However, with great power comes great responsibility. The ethics of modifying life itself is one of the most profound debates of our time.

Understanding the Code of Life

DNA, the blueprint of all living organisms, carries the instructions for life itself. Scientists have deciphered this code, leading to groundbreaking developments such as:

CRISPR Gene Editing: A precise tool that allows scientists to modify DNA, potentially eliminating genetic disorders.

Synthetic Biology: The creation of artificial DNA and even entire organisms designed for specific purposes.

Cloning and Stem Cell Research: The ability to replicate life and regenerate tissues for medical treatment.

These advancements open doors to unprecedented possibilities, but they also raise ethical concerns.

The Promise of Genetic Engineering

Genetic technologies offer immense benefits to humanity, including:

Eliminating Genetic Diseases: Disorders like cystic fibrosis, sickle cell anemia, and Huntington's disease could be eradicated.

Agricultural Advancements: Genetically modified (GM) crops can improve food security by enhancing resistance to pests and extreme climates.

Longevity and Human Enhancement: Scientists explore ways to slow aging and enhance human abilities, from intelligence to physical strength.

However, altering the genetic fabric of life comes with significant moral and societal questions.

The Ethical Dilemmas

1. Playing God: Should We Edit Life?

Altering DNA fundamentally changes an organism, raising philosophical and religious concerns. Is it ethical for humans to modify the natural blueprint of life? Some argue that we are merely continuing evolution, while others warn of unintended consequences.

2. Designer Babies: The Slippery Slope

Genetic editing could allow parents to select traits such as intelligence, height, or eye color for their children. This raises issues of inequality, genetic discrimination, and the commercialization of human life. Should genetic enhancements be limited to medical necessity, or do humans have the right to "upgrade" themselves?

3. Unintended Consequences: The Unknown Risks

Genetic Errors: Modifications could have unforeseen side effects, leading to new diseases or ecological disruptions.

Biodiversity Impact: Genetically altered species, such as engineered crops or lab-created organisms, could disrupt ecosystems.

Without strict oversight, genetic engineering could create irreversible harm.

4. Who Controls the Code of Life?

Biotechnology is largely controlled by corporations and governments. This raises concerns about:

Patents on Genes: Should private companies be allowed to own genetic sequences?

Access to Treatment: Will only the wealthy benefit from life-saving genetic advancements?

Biological Weapons: Could genetic engineering be used for harmful purposes, such as biowarfare?

Regulation and ethical guidelines must ensure that genetic technologies serve humanity rather than exploit it.

Finding a Moral Balance

To navigate these ethical challenges, scientists, ethicists, policymakers, and the public must collaborate. Key steps include:

Regulating Genetic Research: Establishing global standards for safe and ethical genetic modifications.

Ensuring Fair Access: Making genetic therapies available to all, not just the privileged few.

Promoting Public Dialogue: Engaging society in discussions about the moral implications of genetic engineering.

The Future of Life Itself

The ability to edit the code of life presents both incredible opportunities and profound ethical challenges. If guided by wisdom, responsibility, and compassion, genetic science can be a force for good—eliminating suffering and improving quality of life. However, if misused, it could lead to inequality, ethical dilemmas, and irreversible consequences.

As we stand at this crossroads, the question remains: How far should we go in rewriting the fundamental code of life? The future is in our hands, and the choices we make today will shape the destiny of generations to come.

5. A Reflection on Our Place in the Universe

As we unlock the mysteries of the **genetic code**, we gain not only the power to manipulate life but also a deeper understanding of **our place in the universe**. The study of DNA and evolution reminds us that we are part of a **vast, interconnected web of life**, shaped by forces that transcend time, space, and species. Yet, as we wield this new power, we must also recognize that with it comes a **new level of responsibility**—one that requires us to tread carefully, with respect for life's diversity and the future of the planet.

In the grand scheme of the cosmos, our understanding of **DNA and evolution** may be just the beginning of a much **larger story**. As we move forward, we must ask not just how we can manipulate life but **how we can coexist with the living world**, ensuring that our actions contribute to the flourishing of life on Earth and beyond.

Gazing at the night sky, one cannot help but feel a sense of awe and wonder. The vastness of the universe stretches beyond comprehension, filled with billions of galaxies, each containing countless stars and planets. In this grand cosmic tapestry, where does humanity stand? What is our significance in the face of such an immense reality? These are questions that have shaped human thought for centuries, inspiring philosophy, science, and spirituality.

The Scale of the Universe: A Humbling Perspective

The observable universe is estimated to be 93 billion light-years across, containing at least two trillion galaxies. Our Milky Way is just one of them, home to around 100 billion stars. Earth itself orbits a single, average star—our Sun—on the edge of this vast galaxy. From a purely physical perspective, our planet is an insignificant speck in an incomprehensibly vast cosmos.

However, our ability to reflect on our place in the universe is itself extraordinary. Despite our smallness, we possess the unique capability to understand the laws of nature, explore the cosmos, and ask deep existential questions.

The Search for Meaning in a Vast Cosmos

Does the enormity of the universe make human existence insignificant? Some argue that our fleeting lives are meaningless in the grand scheme of things. Yet, others believe that our awareness, creativity, and pursuit of knowledge give life profound meaning. If anything, our existence proves that the universe is not merely a cold, lifeless void—it is also home to conscious beings capable of love, discovery, and wonder.

Are We Alone? The Quest for Life Beyond Earth

For centuries, humans have wondered whether we are the only intelligent beings in the universe. With the discovery of thousands of exoplanets—some potentially habitable—the possibility of extraterrestrial life has become a serious scientific question.

Microbial Life: Scientists search for microbial life on Mars, Europa, and Enceladus, where liquid water may exist beneath the surface.

SETI (Search for Extraterrestrial Intelligence): Radio telescopes scan the skies for signals from alien civilizations.

The Fermi Paradox: If the universe is so vast, why haven't we encountered other intelligent life? Possible explanations range from the idea that advanced civilizations self-destruct to the possibility that they are avoiding us.

Discovering extraterrestrial life would profoundly reshape our understanding of our place in the cosmos.

The Fragility of Life: A Call for Stewardship

The vastness of the universe also serves as a reminder of Earth's rarity and fragility. Our planet is the only known world teeming with life. The more we understand the cosmos, the more we realize how precious and delicate our existence is. Climate change, environmental destruction, and global conflicts threaten the very planet that sustains us. If we are to have a future, we must act as responsible stewards of our home.

The Future of Humanity: Staying or Leaving Earth?

Some believe our destiny lies beyond Earth. Visionaries like Carl Sagan, Elon Musk, and Stephen Hawking have argued that for humanity to survive in the long term, we must become a multi-planetary species. Colonizing Mars, exploring the outer planets, and eventually venturing to other star systems may be the next great chapter in our cosmic journey.

Others argue that our focus should be on preserving Earth rather than escaping it. After all, even with technological advancements, no other known planet can yet replace our own.

Embracing Our Place in the Universe

While the universe is vast and often indifferent, our ability to question, explore, and create gives our existence profound meaning. Whether we are alone or one of many intelligent species, our responsibility remains the same—to cherish life, seek knowledge, and strive for a better future.

In the end, our place in the universe is not determined by size or power, but by the choices we make and the legacy we leave behind. As we continue to look up at the stars, let us do so with both humility and ambition, knowing that the universe awaits our discovery.

Made in the USA
Monee, IL
11 March 2025